Agrochemicals and Pest Management

The Author

Dr. Sathe Tukaram Vithalrao [M.Sc., Ph.D., Sangit Vishard, IBT (Seri.), F.I.S.E.C., F.S.E.Sc., F.S.L.Sc., F.I.C.C.B., F.S.S.I., FHAs] is presently working as Professor and Head, Department of Zoology, Shivaji University, Kolhapur. He has teaching experience of 29 years in Entomology at University PG department and 15 years in Agrochemicals and Pest Management. He has written 30 books and published 255 research papers in national and international journals of repute. He guided 20 Ph.D. students and completed 6 major research projects (from CSIR, DST, DBT and UGC). He visited Canada (1988), Japan (1988), Thailand (2002, 2004), Spain (2005), France (2005), South Korea (2006) and Nepal (2007) etc. for academic work. He is member of editorial board of eleven prestigious journals. He delivered 35 talks through All India Radio and international conferences and involved in Doordarshan, S.T.V. and B.T.V. programmes on useful and harmful insects. He published more than 35 popular articles in daily newspapers on insects and sericulture. He got several prestigious awards like "Environmentalists of the Year-2003", "Bharat Jyoti", "Jewel of India", "International Gold Star", "Eminent Citizen of India", "Education Acumen", "Best Educationist", "Eminent Scientist of the Year-2008", "Lifetime Education Achievement", "Lifetime Achievement in Zoology (Insect Taxonomy)-2009", Education Leadership-2011, Asia Pacific International Award-2012, Global Education Leadership Award-2013, etc. He is also working as Research and Recognition (RR) Committee member for Pune University, Pune; North Maharashtra University, Jalgaon; Shivaji University, Kolhapur and DBA Marathwada University, Aurangabad. He has been awarded several fellowships from different scientific and academic societies. He is Chairman of Maharashtra District Environmental Centre of NESA.

Agrochemicals and Pest Management

T.V. Sathe
Department of Zoology
Entomology Division, Shivaji University
Kolhapur – 416 004

2025
Daya Publishing House®
A Division of
Astral International Pvt. Ltd.
New Delhi – 110 002

First Impression, 2003
Reprinted 2025
ISBN 978-81-7035-309-6

Published by **Astral International Pvt. Ltd.**
– ISO 9001:2015 Certified Company –
4736/23, Ansari Road, Darya Ganj
New Delhi-110 002
Ph. 011-43549197, 23278134
E-mail: info@astralint.com
Website: www.astralint.com

Digitally Printed at : **Replika Press Pvt. Ltd.**

PREFACE

The advancement of mankind depends to large extent on the advancement of knowledge. The scientific findings help man to live more comfortable, protect our property and natural manageable resources. Today, the nation is emphasising more productive agriculture by transforming ancient to modern agriculture. The annual crop loss due to infliction of pests is more than 6000 crores. Out of 30,000 plant diseases recorded from all over the world, Indian plants show about 5,000 diseases. In USA, annual loss due to diseases is about 4 billion dollars. In India, there is no estimation of such losses, probably it may be more than USA. Man must take every possible step to increase agriculture production as there is a tremendous pressure of population on Indian economy. For minimising crop losses there is need of better understanding of crop pests and diseases and their management through agrochemicals.

In the book 14 chapters have been arranged namely, pesticides, chemosterilants, radiational pest control, pheromonal pest control, hormonal pests control, attractants, repellents, anti-feedants, neuropeptides, fungicide, weedicides, chemical fertilizers, nutritional deficiency and bio-fertilizers. The above aspects have direct application in agriculture and are extremely useful for hoping maximum yield of crops. The book will serve as ready notes for agricultural students both at graduate and post-graduate level. I hope that the students, teachers and farmers will find this book as stimulatory and informative.

T.V. Sathe

CONTENTS

Chapter 1
INSECTICIDES

Pesticides have great importance in pest management. Pesticides are those chemicals which kill the pest species. The pests are destructive and troublesome creatures scattered in different animal groups such as mammals, birds, insects, molluscans, nematodes, etc. Pesticides are classified as Insecticides, Avicides, Nematicides and so on. The following classification of pesticides is based on the effectiveness of chemicals against the specific group of animals.

Pesticides

1. Acaricides – kill mites
2. Algicides – kill algae
3. Avicides – kill or repel birds
4. Bactericides – kill bacteria
5. Fungicides – kill fungi
6. Insecticides – kill insects
7. Herbicides (weedicides) – kill weeds
8. Larvicides – kill larvae
9. Molluscicides – kill snails and slugs
10. Nematicides – kill nematodes
11. Ovicides – destroy or kill eggs
12. Picicides – kill fishes
13. Predicides – kill predators
14. Rodenticides – kill rodents

Insecticides are widely used in insect pest management in agro-ecosystem, forest ecosystem, human habitation, cattle sheds and several other places. Insecticides are classified in four ways based on their toxicity, mode of entry, mode of action and chemical nature.

INSECTICIDE CLASSIFICATION

Based on Mode of Entry

Insecticides

- Stomach poisons
- Contact poisons
- Systemic poisons
- Fumigent poisons

Based on Mode of Action

Insecticides

- Physical poisons
- Protoplasmic poisons
- Respiratory poisons
- Nerve poisons

Based on Formulations

Insecticides

- Dusts
- Granules
- Sprays
- Wettable powder
- Emulsifiable concentrate
- ULV
- Fumigants
- Poison baits
- Arosols

Based on Chemical Nature

Insecticides

Plant Origin

1. Nicotine
2. Nornicotine
3. Purethrum
4. Rotenone
5. Neem Dhareck (Azadirachtin)
6. Sabadilla
7. Ryania
8. Acorus
9. Anabasine

Animal Origin

Inorganic

1. *Arsenic Compounds:* Paris green, lead arsenate etc.
2. *Fluorides* Sodium fluoride, Sodium fluociliate, cryolite, etc.
3. *Sulphur* Elemental sulphur, lime sulphur, etc.
4. *Miscellaneous* Sodium selnate, mercury, comp, formaldehyde, etc.

Organic

2. *Chlorinated Hydrocarbons:* DDT, BHC, Chlordane, Aldrin, Endrin, Dialdrin, Endosulphan, etc.
2. *Organophosphorus comp.* Malathion, Parathion, Rogor, DDVP, Phorae, etc.
3. *Carbomates* Carbaryl, Carbofuron, Baygon, Aldicarb, etc.
4. *Dinitrophenols* DNOC, DNOCHP, etc.
5. *Organothio cynates* Lethane, A-70, Thanite, etc.
6. *Fumigants* CCl4, Methylbromide, etc.

Plant Origin Insecticides

Nicotine ($C_{10}H_{14}N_2$)

Pure form of nicotine is 1-3-(1 methyl-2-pyrrolidyl) pyridine or 1-methyl-2, 3 (pyridyl) pyrrolidine. It is highly toxic to mammals and also toxic to soft bodied sucking insects like aphids, jassids, mealy bugs, thrips, etc. It is derived from tobacco plants *Nicotiana tabacum* and *N. rustica*. The pure form of nicotine is water soluble, colourless and odourless liquid. It is also soluble in most organic solvents and formulated as dust, spray and fumigant. Nicotine is contact and stomach poison which affect ganglia at synapses.

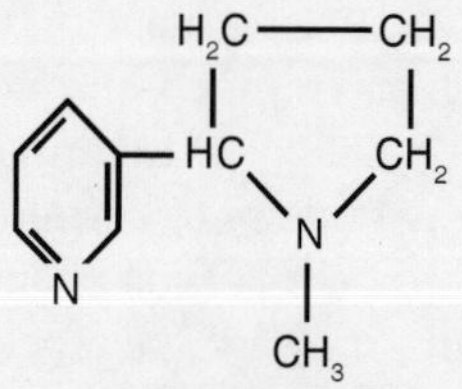

Nicotine

Pyrethrum ($C_{21}H_{30}O_3$)

The pyrethrum was in use against lice and fleas around 1800 in Iran. The pyrethrum is extracted from the white flowers of *Crysanthemum cinerariaefolium*. In pyrethrum, pyrethrum I & II, cinerin I & II and jasmoline I & II are active ingredients as esters. Their structure and formula is similar. These are hydrolysed by alkalis and decomposed by sunlight. Pyrethrum affects the nervous system. It is contact poison. It is effective against aphids, stored grain pests,

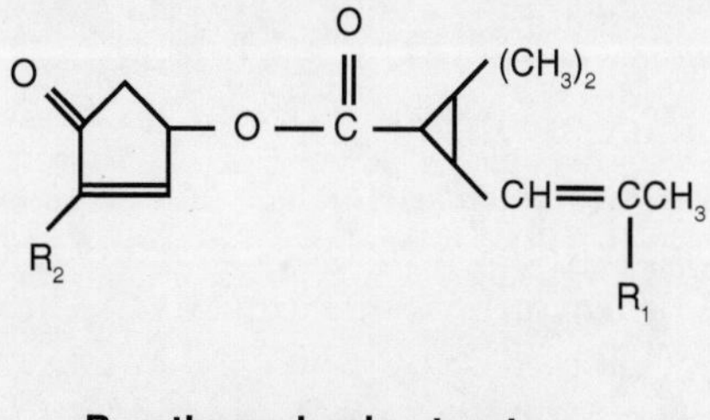

Pyrethrum basic structure

live stock ectoparasites, vegetable pests and several other insects. It shows quick knock down effect and low mammalian and plant toxicolgy. However, it is expensive, and unstable, hence no residual effects observed. The pyrethroids affect both peripheral and central nervous system of target insects which cause repetitive discharge and convulsions in insects. It may block nerve conduction. These pesticides are metabolized by oxidation of M.F.O. Pyrethrums are formulated as sprays and aerosols.

Rotenone ($C_{23}H_{22}O_6$)

It is derived from the plants *Derris* and *Lonchocarpus*. These plants contain 5-9% and 8-10% of rotenone respectively. Rotenone is white crystaline in its pure form. It is rapidly decomposed by alkalis and light. It acts as stomach and contact poison affecting respiration in insects. It is not toxic to mammals and plants. However, it is unstable and has low toxicity to certain insect pests. Rotenone is effective against aphids and other pests of vegetables (Fig. 2) ornamental plants, mulberry and live stock. It controls leaf eating caterpillars.

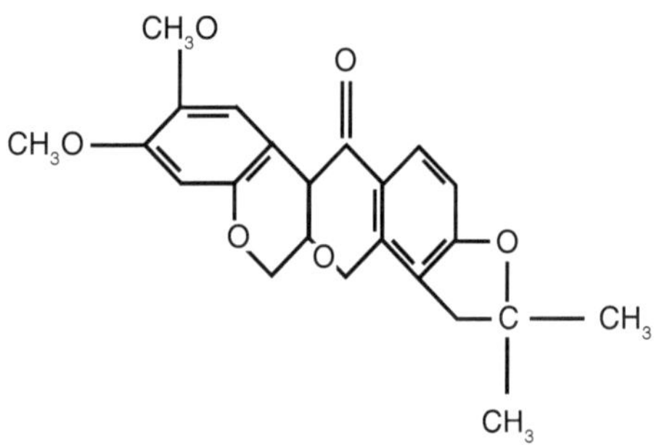

Rotenone

Anabasine (Neonicotine)

It is derived from *Anabasis aphylla* and *N. glauca*. Chemically it is 3-(2-piperdyl) pyridine which is stable and colourless, viscous liquid. It turns brown on exposure to air. Its specific gravity is 1.0481 and B.P. 280-9°C. It turns brown on exposure to air. It is soluble in water and most organic solvents. Nicotinoids affect the ganglia of the central nerve system in insects.

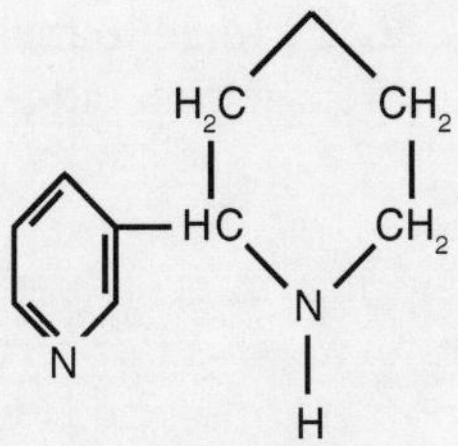

Anabasine

Nornicotine ($C_9H_{12}N_2$)

Nornicotine is found in *Nicotiana sylvestris* and *Duboisia hopwoodi*. Chemically it is 2-(2-pyridyl) pyrrolidine, which is a colourless liquid. It is viscous, hygroscopic and missible in water and organic solvents in all proportions. Nornicotine is more stable and toxic than nicotine. It has B.P. in range of 266-270°C and sp.gr. 1.07 at 20°C.

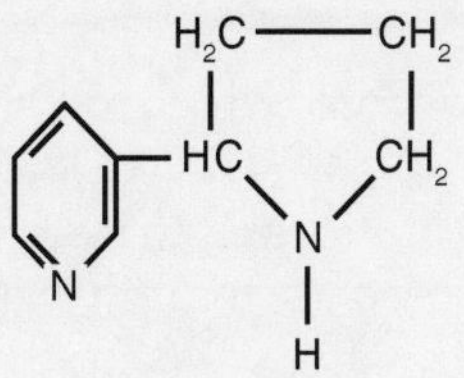

Nornicotine

Sabadilla

Sabadilla is derived from *Sabadila officinale* plant. It is contact and stomach poison, formulated as dusts and sprays. It is a mixture of alkloids veratridine ($C_{36}H_{51}NO_{11}$) and ceradine ($C_{32}H_{49}NO_9$). The seeds of sabadilla contains about 2-4% sabadilla. It probably act on muscle tissues of insects and cause pyralysis. It is effective against hemipterous pests, thrips and garden pests. It also controls lice of human and cattle.

Inorganic Pesticides

As compared to organic and plant origin insecticides, the inorganic insecticides have less importance because they are not species specific and are equally toxic to mammals. Following are the important groups of inorganic pesticides.

Arsenic Compounds

Arsenates leave residues on crop plants and they are less phytotoxic than the arsenites and hence preferred. Some important compounds from this group are described below.

Lead arsenates

The lead arsenates are in basic and acidic forms and were reported first from USA in 1892. Acid lead arsenate ($PbHASO_4$) and basic lead arsenate ($Pb_2(PbOH)(ASO_4)$ are not so useful in pest management. Because, acidic lead arsenate burns the tender parts of the plants and basic one is less toxic to insects. Acid form remains stable under acid conditions (pH 2 to 5) but, changed into basic form under neutral or alkaline condition.

Lead arsenate is available in the form of dust, spray or poison bait ingredient to control locusts, grass hoppers, citrus butterfly, etc. It has very low phyto-toxicity and gives uniform deposits on plants.

Calcium arsenate

It is solid white powder discovered in 1920. It has two forms i.e. Acidic and Basic [$CaHA_5O_4$; $Ca_3(ASO_4)_2$]. These compounds are cheap and comparatively more toxic to insects than lead arsenates. Calcium arsenate can be used as stomach insecticide and as bait. It can control cutworms, grasshoppers, ants, locusts, underground insects, etc. However, this insecticide is phytotoxic. In earlier times it was effective against *Epilachna* beetle on brinjal and potato.

Zinc arsenate

Zinc arsenate is a compound which is effective against certain insect pests.

Paris green

It is bright green powder referred as copper acetoarsenite which was used first in 1865 in USA. Paris green is costly and damage foliage by burning. Hence, it has limited use in pest management. It is a 3:1 or 2:1 mixture of copper meta-arsenite ($3\ Cu\ (ASO_2)_2$) and copper acetate ($(CH_3COO)_2Cu$). It controls slugs and mosquito larvae effectively.

Fluorides

Fluorides are highly active and do not leave highly poisonous residues on the crops.

Sodium fluoride (NaF)

Sodium fluoride is white powder in its pure form. But the marketed insecticidal form is green in colour. It is effective against the ants, cockroaches, lice, etc. It contains 45 per cent of fluorine. It is available in the form of dust. It can be applied with pyrethrins against insect pests. However, it is hazardous to human.

Sodium fluosillicate (Na_2SiF_6)

It is granular, colourless and white powder which is phytotoxic. It is effective against grasshoppers, crickets, weevils, cutworms, etc. It is also toxic to *Bombyx mori* silk worms.

Sodium fluoaluminate (Na_2AlF_6)

It is also referred as cryolite. Its insecticidal properties have been reported in 1929. It is white crystaline form. It can control chewing lice, codling moth, pumpkin beetles, blister beetles. It is one of the safest insecticide from inorganic group. It absorbs the waxy content resulting in loss of water from the insect cuticle and death.

Sulphur

Elemental sulphur

It is available as dust and Emulsifiable concentrate and effective against thrips, mites and certain fungi.

Lime sulphur wash or liquid lime sulphur

The calcium (tetra and penta) sulphides are the active ingredients. Lime sulphur paste is prepared by boiling 100 lb of sulphur and 50 lb of freshly slaked (hydrated) lime, by diluting this can used as spray. It is effective against aphids, scales (Fig. 11), mites, etc.

Miscellaneous Compounds

Bordeaux mixture

It is also reported as "Bordo" which is discovered in 1882 in France. It is mixture of 6 lb copper sulphate, 10 lb lime and 100 lb water. It is a good fungicide and also very good repellent for defoliating insects. It is specially effective against jassids and beetles.

Boric acid (H_3Bo_4)

It is stomach and contact poison and effective against cockroaches and other crawling type of pests found in human habitation. It is safe insecticide but absorbs cutile wax of the insect and cause death.

Sodium selnate (Na_2SeO_4)

It is effective against aphids and mites.

Formaldehyde

An aqueous solution of 40 per cent formaldehyde is a good poison bait for many insects.

Aluminium phosphide

It is effective against root feeders of mulberry.

Mercury compounds

Hg_2cl_2 and $Hgcl_2$ are also good insect pest killers.

Organic Insecticides

Chlorinated Hydrocarbons

Chlorinated hydrocarbons contains chlorine, hydrogen, carbon and occasionally oxygen. They are most famous but useless, pesticides in pest management programmes because, many pests have developed the resistance against several pesticides of this group. Many pesticides are not effective against phytophagus mites and sucking type of insects. Their residues last longer in the environment in active ingredient and thus cause damage to the biocontrol agents and other beneficial insects. The most famous pesticide 'DDT' is from this group only. Some prominent compounds from chlorinated hydrocarbon group are described below:

BHC (Benzene hexachloride) ($C_6H_6Cl_6$)

It is discovered in 1940 as an insecticide and marketed under the trade name "Gammexane". It is contact, stomach and persistent insecticide. In its crude form, it is white to brown product. It is insoluble in water and soluble in organic solvents. BHC contains 6 isomers (Alpha, beta, gamma, delta, eta and epsilon) Gamma isomer is most toxic to insect pests. It is nonphytotoxic and acumulate in the soil as well as in the systems of animals. BHC is toxic to cucurbits and may cause polyploidy. It is formulated as wettable powder, dust, smoke, granules and EC. In gamma isomer metabolism, monodehydrochlorination and gluthion conjugation mechanisms are involved. BHC is effective against all insect pests including mulberry pests like stem girdler, stem borers, termites and several agricultural, forest and household pests.

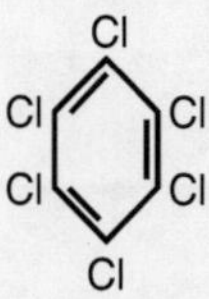

BHC

Toxaphene ($C_{10}H_{10}Cl_8$) (Chorinated terpenes)

Toxaphene is discovered in 1948. It is safer for honey bees and parasitoids but, toxic to mammals. An amber coloured, waxy, solid toxaphene is nonphytotoxic except cucrubits, and formulated as wettable powder, dust, granules and EC. It is effective against grasshoppers, cutworms, jassids and mites.

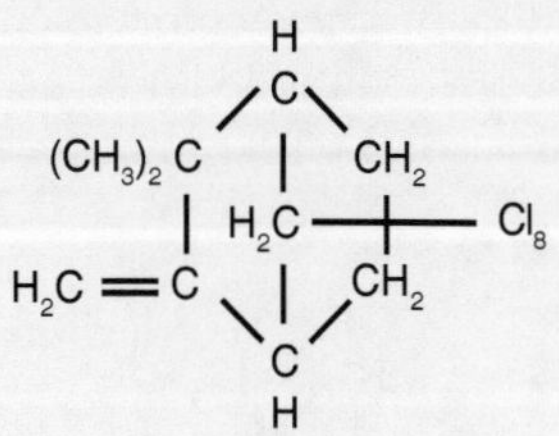

Toxaphane

Chlordane (Cyclodiene group) ($C_{10}H_6Cl_8$)

Chlordane was first used as insecticide in 1945. It is contact and stomach poison insecticide. Its residues last longer in the environment. It is quite effective against earthworms in lawns. Technical chlordane is dark, viscous liquid which is insoluble in water but soluble in most organic solvents. Chemically it is 1, 2, 4, 5, 6, 7, 8-8-octachloro-2, 3, 3a, 4, 7, 7a-hexahydro-4, 7-Methanoindene. It is formulated as 5 per cent dust, 50 per cent WP and 20 per cent EC. It control beetes, weevils and termites very effecitvely. It is also useful insecticide for soil inhabiting pests.

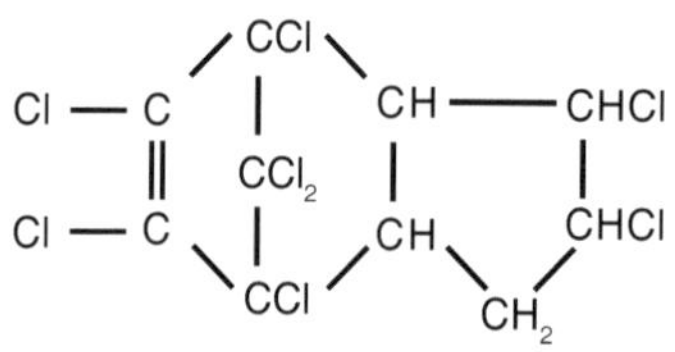

Chlordane

Aldrin ($C_{12}H_8Cl_6$)

It was introduced in 1948 and called as "Octalene". Aldrin is persistent and non-systemic insecticide which is converted into dieldrin through biological oxidation. It is white crystalline compound insoluble in water but, soluble in solvents such as xylene, acetone, Ccl_4, benzene, kerosene, etc. It is formulated as dust and EC and effective against termites, soil insects, grasshoppers, etc. Aldrin is produced by condensing hexachlorocyclopentadiene with the Diels-Alder adduct of cyclopentadiene and acetylene. The pure form is colourless crystalline solid with melting point 104-105°C.

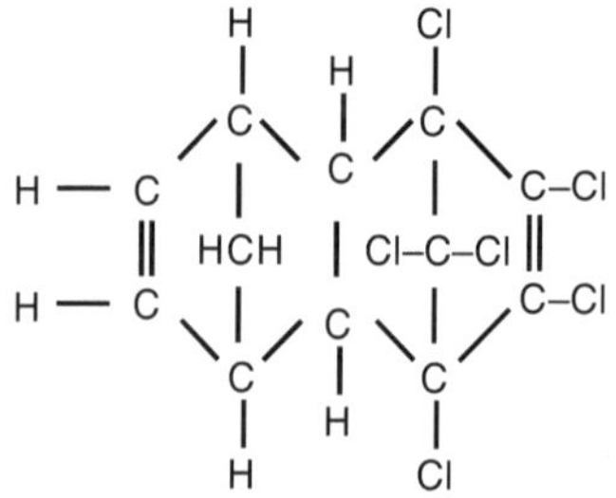

Aldrin

Dieldrin ($C_{12}H_8Cl_6O$)

Dieldrin was first introduced in 1949 as a non-systemic and persistant insecticide. It has very high contact and stomach action. It is also recognized by the common name HEOD. It is formed by the oxidation of aldrin with peracetic or perbenzoic acids. Its pure form is colourless crystalline with melting point of 175-176°C. However, its technical product is buff to light brown. It is highly stable. It has higher dermal toxicity hence, used with care. It is available in the form of WP and EC. It is effective against soil inhabiting insect pests.

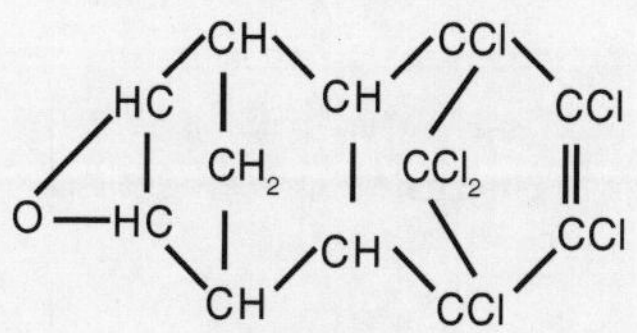

Dieldrin

Endosulfan ($CgH_6Cl_6O_3S$)

Chemically it is heterogenous in structure. It was released in market in 1956 under the trade name, "Thiodan" and is still in use. Endosulfan is both an organic sulphite and chlorinated hydrocarbon. It is brown crystalline solid which is very toxic to mammals and fishes but, it is safer for honey bees and parasitoids of insect pests. It is available in the form of WP, EC, dusts and granules. It is effective against borers, plant bugs and lepidopterous pests [cutworms, *Helicoverpa* sp. (Fig. 17), boll worms (Fig. 16) *Exelastis atomosa* (Fig. 7) and several leaf eaters (Figs. 1-4)]. It also controls aphids, thrips, grasshoppers and several other pests on mulberry

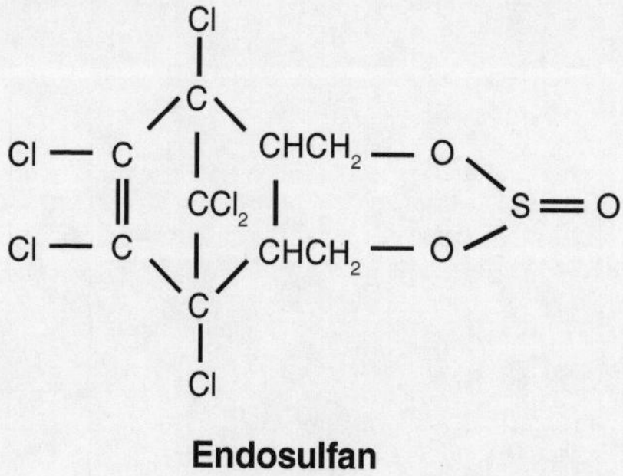

Endosulfan

and other agricultural crops. It is insoluble in water and moderately soluble in most of the organic solvents. It is stable to sunlight. This non-systemic, contact and stomach poison insecticide is metabolized to corresponding sulphate in plants and animals (mammals).

Methoxychlor

Methoxychlor was introduced in 1945 under the trade name "Marlate". However, its insecticidal properties have been investigated in 1944. It is also recognized by different names such as Dimethoxy-DT, DMDT, dianisul trichloroethane. This non-systemic, contact and stomach poison insecticide has also aphicidal and acaricidal activity. Chemically it is 1,1,1-trichloro-2,2-bis (4-methoxyphenyl) ethane. Its pure PP-isomer form is colourless crystal. This safe insecticide is formulated as WP and EC.

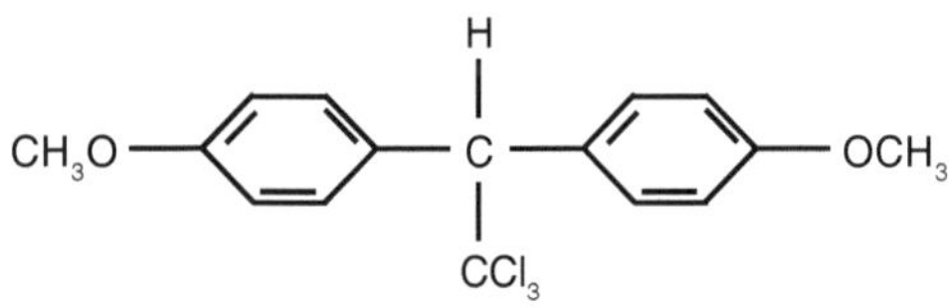

Methoxychlor

Dicofol (Kelthane)

Chemically Dicofol is 2,2,2-trichloro 1,1-bis (4-chlorophenyl) ethanol. This non-systemic insecticide has good acaricide action. Its technical product is brown and insoluble in water but, soluble in most aliphatic and aromatic solvents. It prevents the conduction of impulses due to imbalance of sodium and potassium ions around the neuron.

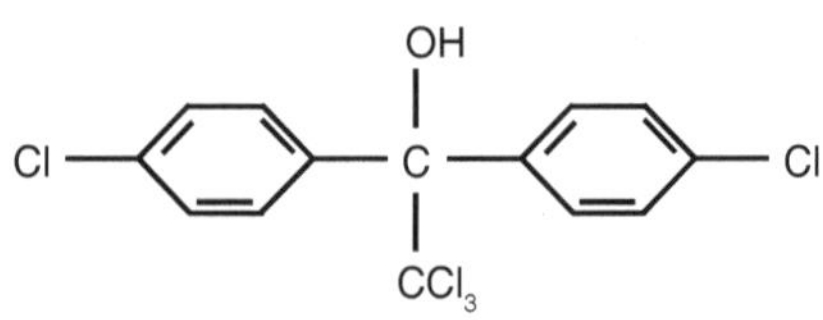

Dicofol

Organophosphorus Compounds

Scharder investigated the first organophosphorus insecticide in 1942. The organophosphorus (OP) compounds contain one or more atoms of carbon, phosphorus, nitrogen, oxygen and sulphur. They have less residual toxicity. However, they are highly toxic to warm blooded animals. Owing to rapid biodegradibility and shorter persistence in the environment, these pesticides have great importance in agroecosystem for controlling varieties of agricultural pests, including pests of mulberry. Most of the pesticides of this group are systemic in action and hence can be safely used against the target species. The biological activities are due to the capacity of the central patom to phosphorylate the active site of the enzyme, cholinesterase (CHE) which is an essential constituent of the nervous system (Singh *et al.*, 1989). Due to the irreversible inhibition of phosphoretated enzyme and accumulation of acetylcholine and consequent disruption of normal functioning of the nervous sytem, a typic cholinergic symptoms, like hyper-sensitivity, hyperactivity, tremors, convulsions, paralysis and death are seen in insects. For degradation of pesticides, a variety of interrelated factors (physical, chemical, microorganisms, etc.) are responsible. The physico-chemical relations may incomplete the degradation, in certain cases the degradation is completed by microorganisms. Monocrotophos and dichlorvos when mixed with fertilizers degrade faster than the phorate and disulfothion. Likely, in all cases of pesticides the increased temperature enhances the process of degradation of pesticides.

Malathion ($C_{10}H_{19}O_6PS_2$)

Malathion, a contact and stomach poison was introduced as an insecticide in 1950 in USA. It is light yellow to dark brown liquid

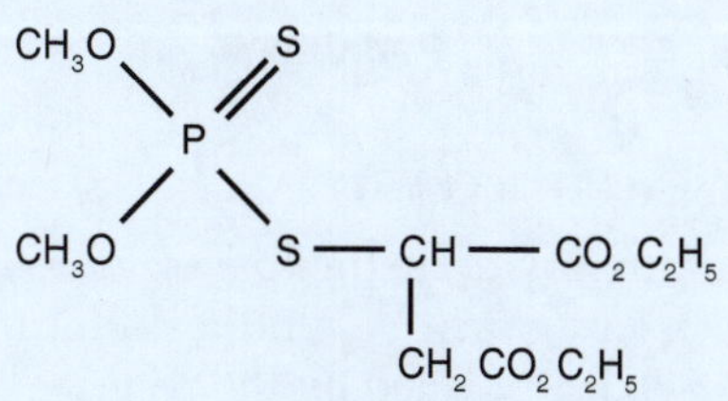

Malathion

which is soluble in water and most organic solvents and mineral oils. It has low phytotoxicity and strong offensive odour and unstable in an alkaline medium. Due to its less mammalian toxicity it is safer for use against the pests of edible crops, live stock, household, storage, mulberry and ornamental pests. It is moderately persistent pesticide which is quite effective against scales, jassids, mealy bugs, weevils and stem borers. It is widely used for controlling caterpillers, aphids, jassids, psyllids plant bugs etc. Malathion is converted into malaoxon by MFO and further by carboxylesterase to malathion carboxylic monoacid which is nontoxic and excreted. Malathion is formulated as dust, pure liquid and EC.

Parathion ($C_{10}H_{14}NO_5PS$)

Parathion was introduced by the American Cynamid Co. under the trade name of "Thiophos" in 1947 and "Folidol" in 1948 by Bayer. It is non-systemic, contact and stomach insecticide with some fumigant action. It is nonphytotoxic and effective against scale insects, aphids and mites. However, it is phytotoxic to some ornamental plants and horticultural plants. Its residues degrade shortly in environment but toxic to mammals hence, not recommended against household and live stock pests. Its use should also be avoided on blossoming crops as it causes damage to flowers and secondly it is responsible for killing bees and other pollinators of the crops. The formulations include WP, EC, dust, smoke and aerosol. It function by inhibiting the acetyl cholinestearase.

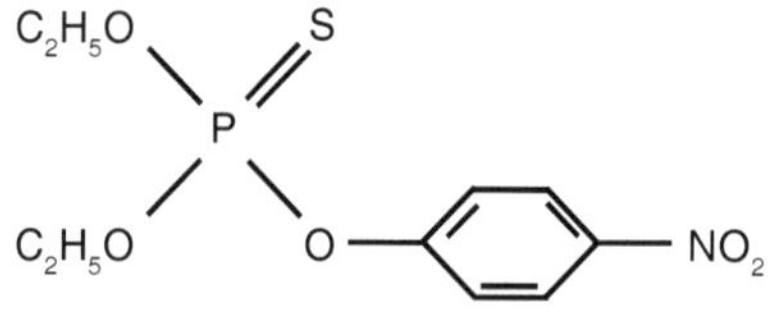

Parathion

Methyl parathion ($C_8H_{10}NO_5PS$)

Mammalian toxicity of methyl parathion is considerably low hence, widely used against several agricultural pests. It was introduced in 1949. It is formed by the reaction of O, O-dimethyl phosphoro-chloridothioate with sodium 4-nitrophenonide. It is soluble in most of the organic solvents. Its pure form is colourless crystalline powder which is soluble in 55-60 mg/l water. It is formulated as dust and EC and effective against sucking insects (Figs. 10-11) and foliage feeders (insects) (Figs. 1 & 4).

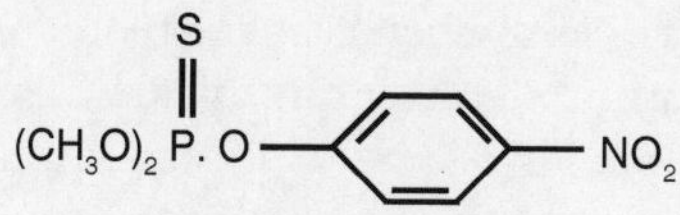

Parathion-methyl

Rogor (Diamethoate) ($C_5H_{12}O_3PS_2N$)

Rogor was first introduced in 1956 in Italy and USA. Initially its use was restricted to fruit flies. It is white crystalline solid, unstable in an alkaline medium. It has light colour and soluble in solvents like alcohol, acetone, benzene, chloroform, etc. and less soluble in water. Rogor is contact and systemic poison and phytotoxic to some plants like Begonia, Figs, etc. Its mammalian toxicity is very low. Diamethoate is specially effective against live stock pests and sucking type of pests. But, it is destructive to bees in the field. It can control the psyllids very effectively present on mulberry, forest trees and other plants with 0.09 per cent spray. It is equally effective against psyllid galls, chewing and sucking insects specially jassids, aphids, etc. and cattle grubs. Rogor can be produced by 3 ways:

(1) Reacting sodium O, O-dimethyl phosphorodithioate with 2-chloro-N-methylacetamide.

(2) Treating with methylamine and sodium O, O-dimethyl phosphorodithioate and phenyl chloroacetate.

(3) Treating methylamine with methyl (dimethoxy phosphino-thioylthio) acetate.

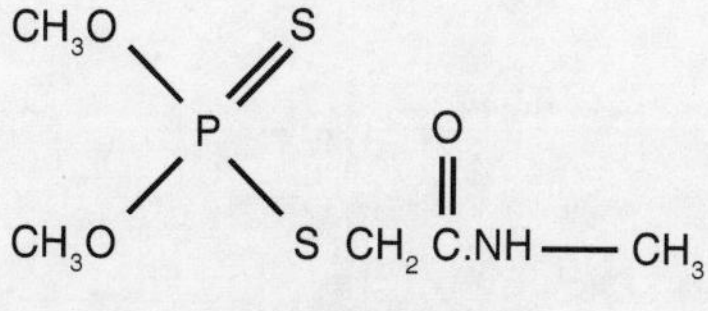

Rogor

Phorate ($C_7H_{17}O_2PS_3$)

It was introduced in 1954 as "Thimet". It is useful for treating soil and seeds at sowing time. Phorate colourless liquid which is soluble in many solvents like alcohol, CCl4, xylene, etc. But, it is less soluble in water. It does not persist in the plant for long, and oxidized to sulfone. It is formulated as grannules, EC and dusts. Phorate is effective against sucking type of insects, mites and nematodes. Phorate is contact, systemic and slight fumigant in action and unstable in humid conditions and alkaline medium.

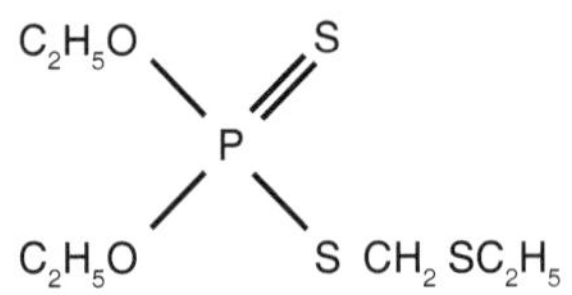

Phorate

Diazinon ($C_{12}H_{21}N_2O_3PS$)

Diazinon is non-systemic contact poison which was introduced first in 1952 under the trade name "Basudin". It is non-phytotoxic and safer than parathion. It control aphids, thrips, flies and mites very effectively. In irrigated crop system it gets translocated in the plant through capillary action and thus, kill tissue borers feeding internally. It can be produced by condensing isobutyramie with ethyl 3-oxobutyrate and treating resulting pyrimidine with O, O-diethyl phosphoro-chloridothioate.

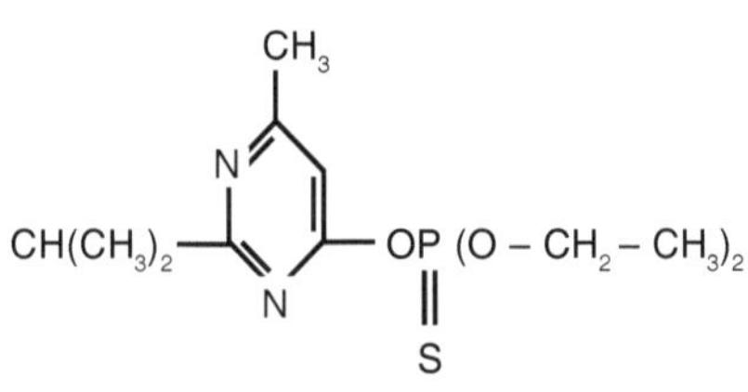

Diazinon

Chloropyriphos ($C_9H_{11}Cl_3No_3PS$)

Its pure form is colourless crystallines. It can be produced by reacting 3,5,6-trichlorpuridin-2-01 with O, O-diethyl phosphorochloridothioate. It is readily soluble in most of the organic solvents and water (at 2 mg/l). This non-systemic contact insecticide is effective against household pests, many agricultural pests and ectoparasites of cattles. It is also effective against termites and white grubs when applied to the seeds.

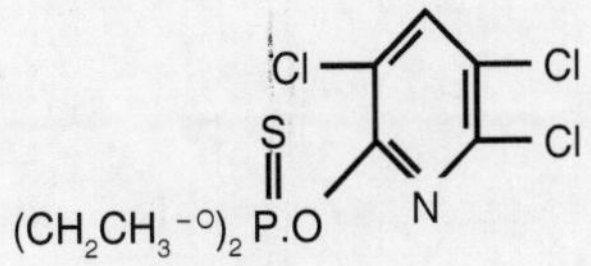

Chloropyrifos

Demeton ($C_8H_{19}O_3PS_2$)

It is available in market under the trade name "Systox". It is soluble in organic solvents and in water at the extent of 60 mg/l (d.o.). It has two forms, Demeton-O and Demeton-S. Its technical product is light yellow liquid. This systemic insecticide acts as acaricide and very effective against sucking insects.

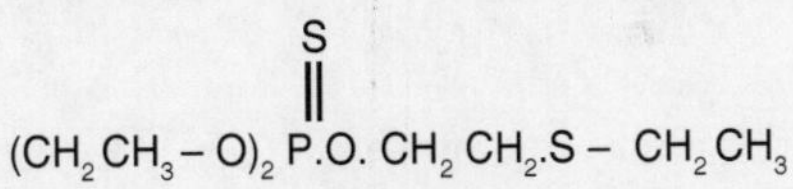

Demeton

Asinphos methyl

It was first introduced in 1953 under the trade names "Gusathion-M" and "Guthion". It acts as contact poison, non-systemic insecticide, acaricide and ovicide and effective against many insects. It is available as WP, EC and dust.

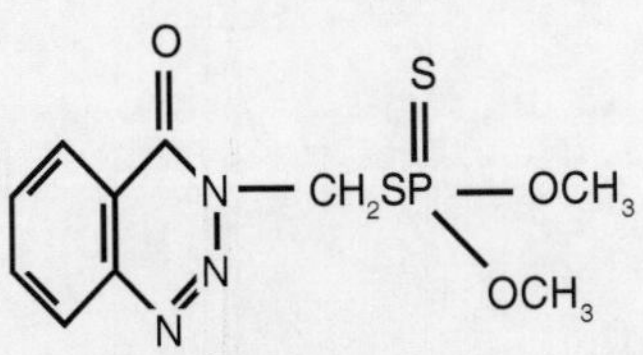

Gusathion

Dichlorvus (DDVP) ($C_4H_7O_4Cl_2P$)

It was introduced in 1955 initially for controlling house flies. DDVP is pleasant smelling liquid. It is stomach, contact and fumigant in action which shows quick knock down effect in insect pests. It is soluble in water and organic solvents and unstable in alkaline medium but it is non-phytotoxic and non-persistent. It is toxic to bees and formulated as EC. This insecticide is useful for controlling many agricultural and non-agricultural pests such as cut worms, *Spilosoma* (Fig. 12), wasp moth caterpillar, *Acherontia* and may other caterpillars; jassids, scales, mealy bugs, thrips, leaf miners, leaf folders, many defoliators. It can be produced by condensing trimethyl phosphite with trichloroacetaldehyde or by dehydrochlorination of trichlorophon. Recent formulations are granules and aerosols.

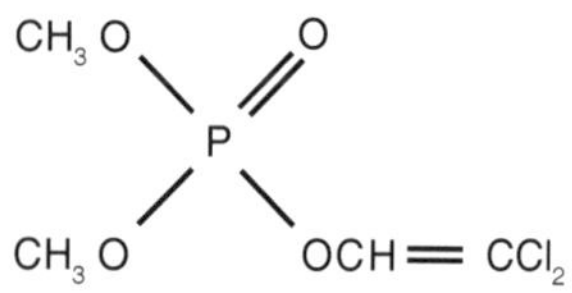

Dichlorvos

Metasystox R' (Oxydemeton-methyl) ($C_6H_{15}O_4PS_2$)

Metasystox was first introduced in 1945 which is contact and systemic insecticide and acaricide. It is yellowish to colourless liquid,

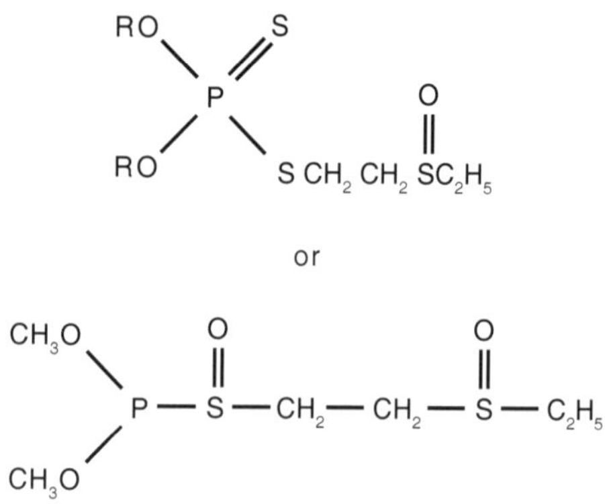

Metasystox

soluble in most organic solvents and unstable in alkaline medium. It is metabolized to sulphoxide and sulphone in the plants. It controls semiloopers and leaf miners.

Monocrotophos ($C_7H_{14}O_5NP$)

It is referred as "Nuacron". This systemic insecticide is less soluble in xylene and least in kerosene but, miscible in water. It is white crystalline solid which rapidly dydrolized by alkali. It is effective against mites, aphids, thrips, etc. It is also effective against stem borers and mulberry pests such as thrips, jassids, scales, mealy bugs etc.

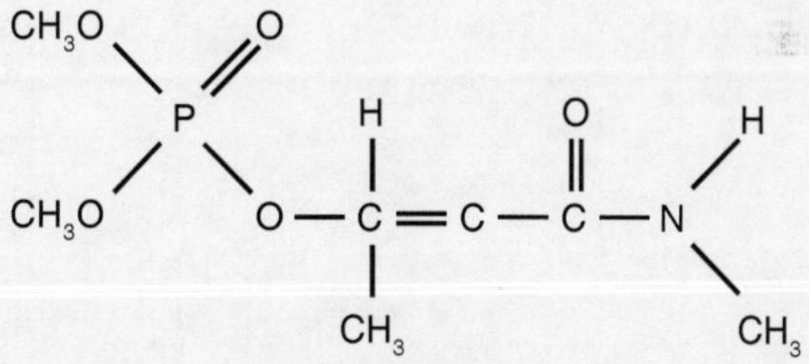

Monocrotophos

Acephate ($C_4H_{10}NO_3PS$)

Its registered trade name is "Orthene" (Cc comp). Its insecticidal action is examined by Grayson in 1972. Its pure form is crystalline solid and white. It has pungent rotton cabbage like odour. It melts at about 82-89°C and it has specific gravity of 1.35. It is non-phytotoxic and systemic insecticide. It can be produced by the acetylation of O, O-dimethyl phosphoramidothioate followed by isomerisation to the O, S-dimethyl analogue. It is available in the form of soluble powder, and effective against pests on cotton, brinjal, cabbage, etc. It effectively

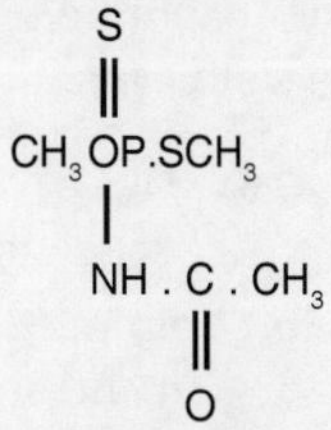

Acephate

control jassids, red cotton bug, boll worms on cotton and jassids and borers on brinjal. It is safer for wildlife.

Phosalone ($C_{12}H_{15}ClNO_4PS_2$)

Chemically it is S-6-chloro-2,3-dihydro-2-oxobenzoxazol-3-ylmethyl O, O-diethyl phosphorodithioate. Its technical grade is colourless solid with garlic odour. It has melting point, 48°C. It is soluble in benzene, acetone, chloroform, methanol and 100 ml/l water. It is very good insecticide and acaricide which is effective against soft bodied insects and mites. It is popular under the trade name "Zolone". It is produced by the condensation of sodium O, O-diethylphosphorodithioate with 6-chloro-3-chloromethyl-3H-benzoxazol-2-one.

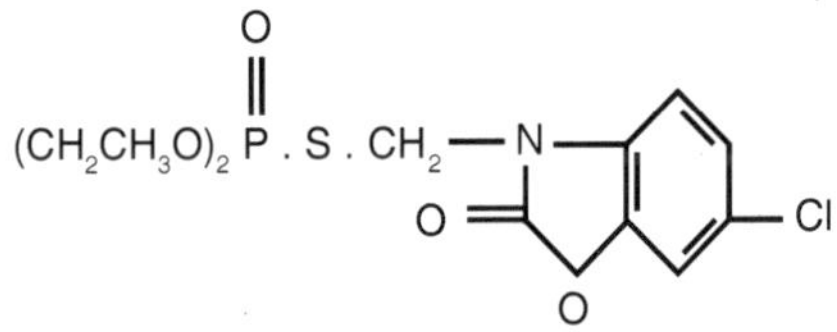

Phosalone

Profenofos ($C_{11}H_{15}BrClO_3PS$)

It is released by Ciba Geigy AG under the trade name "Curacron". Chemically it is O-(4-bromo-2-chlorophenyl) O-ethyl S-propyl phosphorothioate. It is highly toxic to birds and fishes. It is soluble in most organic solvents.

It is produced by the reacting 4-bromo-2-chlorophenol with O-ethyl S-propylphosphoro-chloridothioate. Its technical form is pale yellow liquid. It is soluble in water (20 mg/l) at 20°C. This non-systemic insecticide is recommended against the boll worm complex of cotton, and pests of vegetables.

Phosphamidon (Dimecron) ($C_{10}H_{19}O_5NClP$)

Dimecron was introduced in 1955 as systemic and contact insecticide and acaricide. Its commercial product is a violet liquid but, in its technical form is brown and soluble in most solvents and miscible in water. Its mammalian toxicity is moderate hence, may be used for controlling pests on edible crops. It has very wide range of pests for control. It is available in the form of WP.

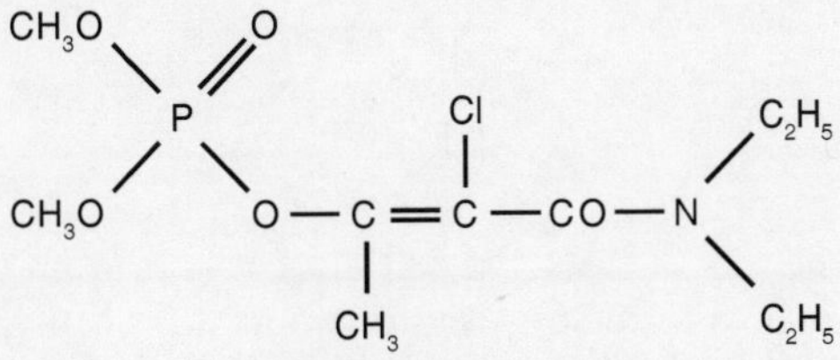

Dimecron

Monocrotophos ($C_7H_{14}O_5NP$)

It is referred as "Nuacron". This systemic insecticide is less soluble in xylene and least in kerosene but miscible in water. It is white crystalline solid which rapidly hydrolyzed by alkali. It is effective against aphids, thrips, mites, stem borers, scales, mealy bugs, etc.

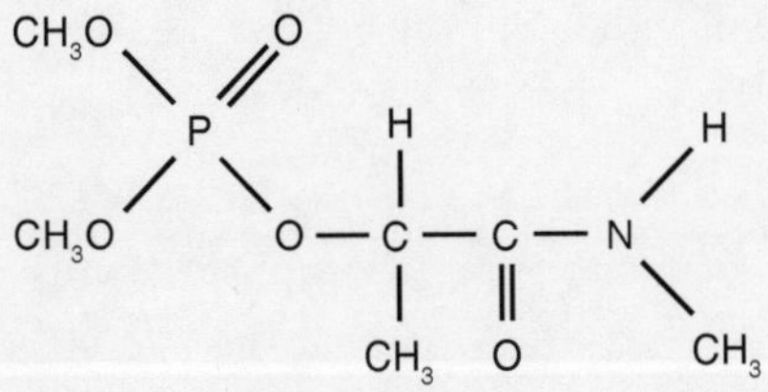

Monocrotophos

Fenetrothion ($C_9H_{12}NO_5PS$)

Fenetrothion is yellowish brown liquid having contact and stomach action, also has good depth action. It is produced by reacting O, O-dimethyl phosphoro-chloridothioate with an alkali metal 3-methyl-4-nitrophenoxide. It controls sucking pests, borers and mites very effectively. Fenitrothrion is available in the form of dust, EC and ULV and can be mixed with any conventional insecticides and fungicides. It is advised that alkali reacting compounds should be used immediately after mixing.

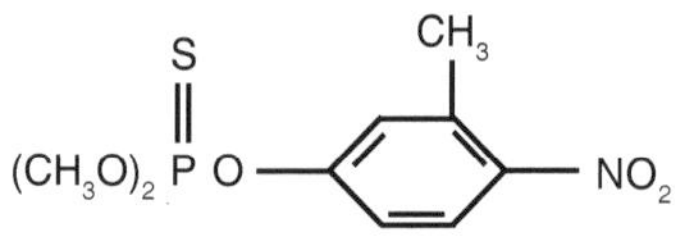

Fenitrothion

Menazon

This potent aphicide was introduced first in 1961 under the trade name "Sayfos". It is non-phytotoxic and effective against aphids and other insects. It is specially used for treating tubers and seeds for protection from insect pests. It is toxic to bees but its mammalian toxicity is quite low. WP and seed dressing suspensions are the commonly used forms of menazon.

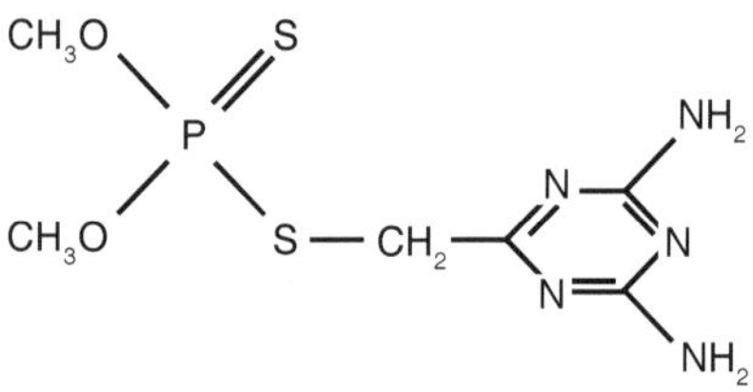

Menazon

Formothion

It was first introduced in 1964 under the trade name "Anthio". It is systemic insecticide with great aphicidal and acaricidal action. It has low mammalian toxicity hence useful to edible crops for pest control. It is specially effective against insect pests found on cotton, jute, rice, wheat, maize, jowar, sugarcane and many more crops.

Fonofos ($C_{10}H_{15}OPS_2$)

Chemically it is (±) O-ethyl S-phenyl ethyl-phosphonodithioate. Its R-isomer is more toxic to insects than S-isomer. Fonofos is effective against soil inhabiting insects such as root worms, wire worms, white grubs, crikets, etc.

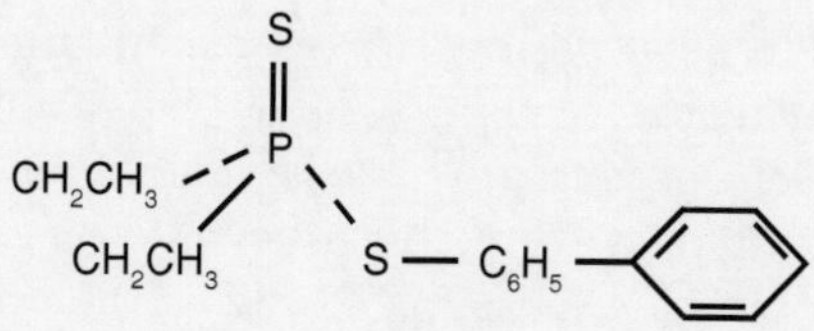

Fonofos

Carbamates

This group of insecticide is relatively new and emerged just before two decades. The carbamates are characterized by having less residual toxicity, low water solubility and organic solvent solubility. These insecticides are derived from carbamic acid containing OCNOCON group which can be represented by the following formula:

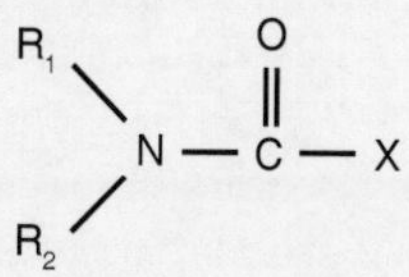

where R represents an alkyl group and X, represents a diversified group.

These insecticides are highly poisonous to insects and other animals and are strong inhibitors of cholinesterase and have direct effect on acetylcholine receptors. They have very rapid penetration capacity into the insect body. Some members of carbamate insecticides are given below. Carbamates have two groups *viz.*, phenyl carbamates and Oxime carbamates.

Carbaryl ($C_{12}H_{11}NO_2$) (PC)

This contact insecticide was introduced in 1956 under the trade name "Sevin". It is biodegradable easily hence, does not accumulate in the ecosystem. Carbaryl has slight systemic action and is non-phytotoxic at insecticidal concentration. It is safer for insect pollinators, parasitoids and predators. Carbary is white crystalline solid, insoluble in water and moderately soluble in xylene, ketone, CCl_4 and petroleum ether, etc. It has broad spectrum thus, effective against the pests of fruits, cotton, vegetables and many other crops. It is compatible with most pesticides except under alkaline conditions. It is available as 50 per cent WP, 5 per cent dust, 4 per cent granules and 85 per cent sprayable powder. It can be produced by reacting 1-naphthol with methyl isocynate or with carbonyl chloride and methyl amine. It has sp.gr. 1.232 at 20°C and M.P. 142°C.

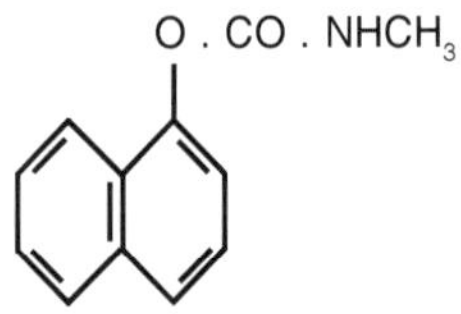

Carbaryl

Carbofuran ($C_{12}H_{15}NO_3$) (PC)

It is systemic in action and soluble in solvents but, slightly soluble in water. This colourless crystalline insecticide was introduced by Bayer in 1967. It has slight phonolic odour and M.P. at 154°C. It is non-phytotoxic compatible with most of the pesticides. It can be produced by the thermal rearrangement and cyclisation of 1-methyl allyloxy-2-nitrobenzene to 2,3-dihydro-2, 2-dimethyl-7-nitronenzofuran, reduced to amine, diazolized and converted to 2,3-dihydro-2, 2-dimethylbenzofuran-7-01, and esterified with methyl isocynate. It is effective against aphids, flies soil inhabiting insect pests, borers and other cell sap suckers.

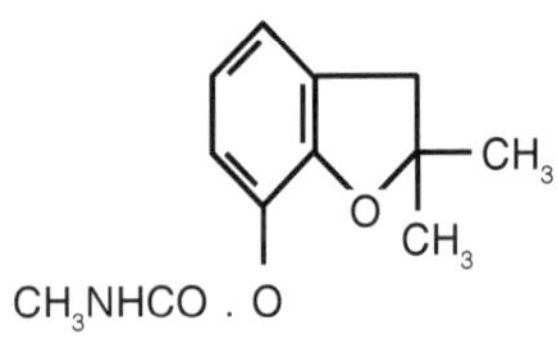

Carbofuran

Amino carb ($C_{11}H_{16}O_2N_2$) (PC)

This stomach insecticide was first released in 1963 under the trade name "Metacil". Chemically it is 4-dimethyl-amino-3-methyl phenyl N-methyl carbamate. It is odourless white crystalline solid which is slightly soluble in water and moderately soluble in aromatic solvents but, readily soluble in polar organic solvents. It is effective molluscicide and insecticide. It controls caterpillars (Figs. 12 & 14).

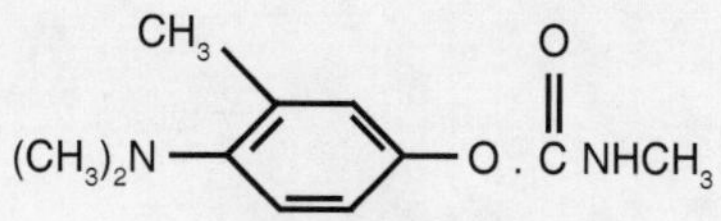

Aminocarb

BPMC ($C_{12}H_{17}NO_2$) (PC)

Its technical product is crystalline solid and less toxic to fishes than many other insecticides. It is quite safer for predatory lady bird beetles and spiders and honey bees. It is non-phytotoxic and soluble in organic solvents such as xylene, acetone, benzene etc. and in water 660 mg/l at 30°C. BPMC can be prepared by reacting 2-sec-butylphenyl chloroformate with methylamine. It is available for use in pest management as 50 per cent EC, 4 per cent dust and 4 per cent granules. It is effective against stem borer (Fig. 6), gall midges, mustard aphids (Fig. 10) jassids on mango and pyrilla on sugarcane.

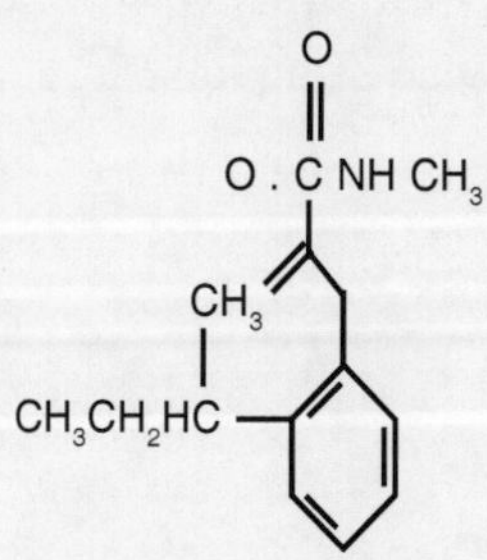

BPMC

Bendicarb ($C_{11}H_{13}NO_4$) (PC)

It was first introduced in 1971 under the trade name "Ficam". Its pure form is colourless solid. Chemically it is 2, 3-isopropylidene-dioxyphenyl methyl carbamate. It is contact and stomach insecticide which is soluble in hexane, acetone, ethanol and water (40 mg/l). Bendicarb is prepared by reacting 2, 3-isoprophylidene dioxyphenol with methylisocynate. It is available in the form of WP, granules and dust and effectively controls household pests such as houseflies, mosquitoes, sandflies, ants and cockroaches. It is specially effective against lepidoteran pests, coleopterans, and collumbollans and soil insects.

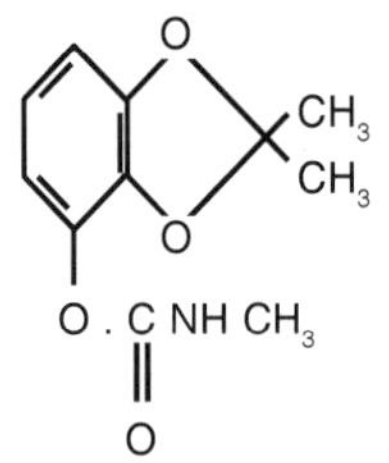

Bendicarb

Aldicarb ($C_7H_{14}N_2O_2S$) (OC)

Aldicarb was released in the market in 1965 under the trade name "Temik". It is colourless crystal with melting point 98-100°C. It is soluble in water (6 g/l) and in most organic solvents. This systemic insecticide can be produced by reacting methylisocynate with 2-methyl-2-(methylthio) propionaldehyde. Chemically aldicarb is 2-methyl-2 (methylthio) propionaldehyde O-methyl-carbamoyloxime. It is effective against mites, insect pests and nematodes.

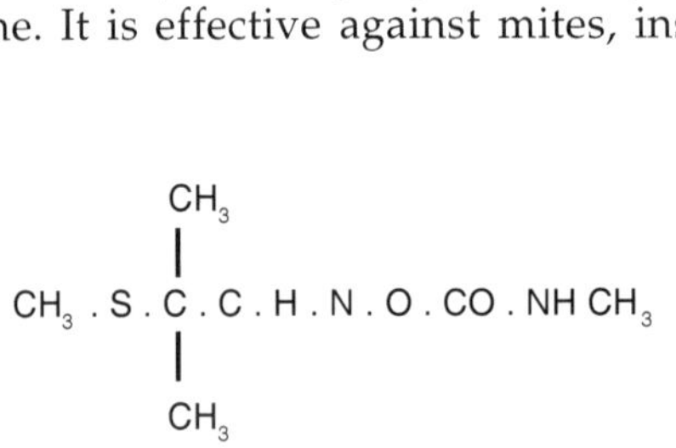

Aldicarb

Methomyl ($C_5H_{10}N_2O_2S$)

Its technical grade is crystalline solid colourless with sulphurous odour. It was introduced first in 1967 under the trade name "Lannate". It is contact insecticide and effective against a large number of pests such as armyworm, cabbage semilooper, fruit borers, leaf defoliators, cotton boll worm, okra stem fly and other borers.

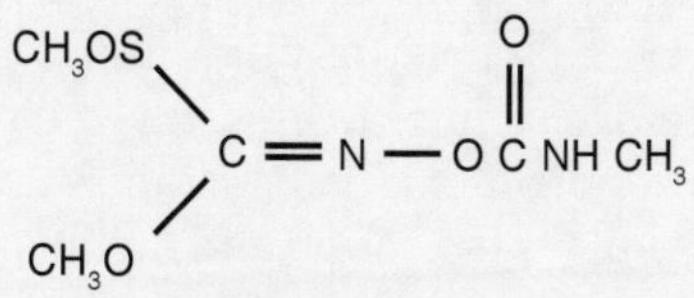

Methomyl

Isolan

It has short residual action. It is colourless liquid which is soluble in water and solvents and effective against aphids and flies.

Other members

Baygon, Oxamyl, Aldoxycarb, etc.

Dinitrophenols

These insecticides has high phytotoxicity hence not accepted for pest management on large scale. Dinitrophenols are contact and stomach poisons. These are derivatives of 4, 6 dinitro-2-alkyphenols and their salts or esters. These insecticides interacts with insects by preventing utility of nutritional energy by blocking oxidative phosphorylation and inducing rapid weight loss.

Dinitro-O-cresol (DNOC)

It was released in 1982 under the trade name "Antinonin". It is colourless, yellowish solid insecticide which is insoluble in water. It is a good acaricide, herbicide and fungicide. It has also ovicidal action for insects and mites. It is effective against locusts and grasshoppers. It controls powder mildew and broad leaved weeds.

Dinoseb

It is effective against mites and insects on fruits. It is used as a spray.

Binapacryl

It is acaricide, useful in agro-ecosystem for controlling mites.

Organo Thiocyanates

Organothiocyonates have low toxicity towards warm blooded animals and are rarely phytotoxic. Good number of compounds are available in market, which are highly toxic to several insect pest species. They are highly toxic to aphids, leaf hoppers (jassids), mealy bugs, thrips, etc. These insecticides have creosote like odour and safe for human and other domestic animals. These compounds interfere with cellular respiration and metabolism and cause death of the insects. They also act on cytochrome oxidase. Thiocyanates is an important component in aerosols. Lethane 60(R), Lethane 384(R), Lethane A-70 and Thanite(R) are good examples.

Organo Sulfurs

They have sulfur as the central atom on which remaining molecules are attached. These are effective against vegetable and fruit pests (mites).

Tetradifon ($C_{12}H_6Cl_4O_2S$)

It was introduced first in 1954 under the trade name "Tedion". It is colourless crystalline solid material with melting point 148°C. It is soluble in water (200 mg/kg). Chemically it is 4-chlorophenyl 2,4,5-trichlorophenyl sulphone. It is non-systemic with acaricide action. It effectively controls phytophagus mites attacking vegetables, fruits, ornamental plants and cotton.

Tetrasul ($C_{12}H_6Cl_4S$)

It is colourless crystalline solid with M.P. 87.5°C. Chemically it is 4-chlorophenyl 2,4,5-trichlorophenyl sulphide. It was released in the market in 1957. It is slightly soluble in water, moderately in acetone and readily in benzene and chloroform. This non-systemic acaricide is also effective against rats. Its technical product is yellowish brown to cream solid.

Fumigants

Fumigants are chemicals containing organic molecules, one or more of the halogens Cl, Br or F employed in gaseous state for controlling insects or other pests. Fumigants are volatile and normally heavier than air and are highly penetrating. Fumigants are useful to control pests in closed environment effectively. They are also used in open for controlling several pests. Fumigation is the best control for many pests species to adopt in store houses, godowns, empty ware houses, ships, flour mills, libraries etc.

Choice of fumigants

Fumigants can be selected by considering following aspects:

1. Cost of material and application
2. Toxicity to insects, plants and other animals
3. Safety to operator
4. Possibility of the fire hazard
5. Bleaching effects to household items
6. Persistence
7. Penetration power
8. Germination of seeds
9. Tainting or preserving qualities of food
10. Known antidote and treatment

However, fumigant efficacy depends on temperature, humidity, concentration, evaporation, diffusion, penetration, sorption, pressure, exposure period, load factor, chemical nature and particle size, etc.

Carbon disulphide (CS_2)

It is pale yellowish to colourless liquid with boiling point 46.3°C and melting point 108.6. It is miscible in ethanol, chloroform and diethyl ether. CS_2 is highly flammable and has unpleasant odour. It is used in mixture with carbontetrachloride against stored grain pests. It is also effective against nursery stock pests and nematodes. For treating clothing, fabrics, seeds, cereals, it is recommended with 1 pound per 100 cubic feet.

Carbon tetrachloride (CCl_4)

It is colourless liquid with boiling point 76°C and melting point 23°C and vapour pressure 114.5 mm Hg at 25°C. It is non-explosive, non-inflammable and soluble in water (280 mg/kg at 25°C). It is used with CS_2 or ethylene dibromide etc. as it has low insecticidal activity, and to reduce fire hazards. It is useful for controlling helminthes parasites of live stock.

Ethylene dichloride ($C_2H_4Cl_2$)

Ethylene dichloride is colourless liquid, non-corrosive,

flammable and with chloroform like odour. It is soluble in most organic solvents and water (4.3 g/l). It has B.P. 83.5°C, M.P. 36°C and VP 78 mm Hg at 20°C. Ethylene dichloride is effective against stored grain insect pests when used in mixture with carbon tetra chloride.

Methyl bromide (CH_3Br)

It is colourless gas, with B.P. 3.6-4.5°C. It is soluble in water (13.4 g/kg) and most organic solvents, forming liquid which is highly toxic to man. It has very high penetrability and insecticidal action. It also acts as weedicide, fungicide, nematicide and acaricide. It is available in the market as glass ampules, metal cans and cylinders. It is used with 1-3 pound per 1000 cubic feet with an exposure of 12 to 24 hours. It is effective fumigant for killing the pests of tubers, bulbs, flour, clothes, etc. It controls soil pests and green house pests.

EDCT (Mixture of ethylene dichloride - 3 parts and carbon tetrachloride -1 part)

It is not suited for milled products. It has several important features which include, it is least hazardous, slow insect killer and slow vaporizer, exposure period is from 24-72 hours, diffusion is slow, required prolonged ventilation after treatment.

Hydrogen Cyanide (HCN)

It is colourless liquid, with almond like odour, MP 14°C and BP 26°C. It is very toxic to man but used for fumigation of stored grains, flour mills, etc. It damages respiratory organs and nerve centres causing death. It is lighter than air and widely used for treating green houses, grain mills, ships, storage ware houses, citrus trees and dwellings.

Ethylene dibromide ($C_2H_4Br_2$)

It is colourless liquid with BP 131.5°C, MP 9.3°C, VP 11.0 mm Hg at 25°C. Chemically it is 1,2-dibromoethane. It is soluble in water (4.3 g/kg), and most organic solvents. It is non-flammable and effective against stored grain pests, fruit pests and vegetable pests and also soil nematodes.

Chloropicrin (CCl_3NO_2)

Chloropicrin is colourless liquid with BP 112°C, MP 64°C and VP 23.8 mm Hg at 25°C. It is non-corrosive to copper, bronze and brass but corrosive to iron, zinc, etc. It is soluble in water (2.77 g/l). It is effective against soil nematodes and soil fungi.

Aluminium phosphide (AlP)

It was introduced in 1930. It appears in yellowish dark crystals which on exposure to moist air produces toxic phosphine gas. It has ammonical odour and BP 87.4°C. This fumigant is used at the rate of 1-3 tablets/tonne grain. It is available as phostoxin pellets, celphos (Excel industries). It is effective against rats. It is also effective to all stages of insects but no adverse effect on seed viability.

Other fumigants used in insect pest management includes:

1. Carbon dioxide (CO_2)
2. Ethylene oxide $(CH_2)_2O$
3. Paradichlorobenzene ($C_6H_4Cl_{12}O_2$PDB)
4. Orthodichloro benzene ($C_2H_4C_{12}$)
5. Naphthalene ($C_{10}H_8$)
6. Nicotine ($C_{10}H_{14}N_4$)
7. DD mixture
8. Azobenzene ($C_{12}H_{10}N_2$)

Synthetic Pyrethroids

Recent synthetic pyrethroids have attracted the attention of farmers and entomologist because some of the new members are very effective against a variety of insect pests. However, the use of pyrethrum was not popularised because of high production cost, instability in sunlight and uncertainty of availability. On the contrast, synthetic pyrethroids shows many desirable attributes which are given below:

1. They have extremely high insecticidal activity at extremely low doses.
2. They are biodegradable.
3. They affect almost all stages of the pests.
4. They are not readily washed off from the crops since they are lipophlic.
5. They have low volatility and polarity.
6. Mammalian toxicity is low.

Allethrin was the first synthetic pyrethroid released in the market in 1949. Later, many inefficient compounds were investigated and released in the market. However, only few, permethrin, fenvlerate, cypermethrin, etc. proved their worth in pest management.

Permethrin ($C_{21}H_{20}Cl_2O_3$)

It was investigated in 1973 and marketed under the trade names, "Permosect" and "Ambush". It is contact insecticide. Chemically it is methyl (3-phenoxyphenyl) ± cistrans, 3-(2,2-dichlorvinyl)-2, 2-dimethyl cyclopropane carboxylate. It has molecular weight, 391.3. Its technical form is viscous, liquid or semi solid, colourless to pale brown, and with BP of 200°C, MP 35°C (approximate) and VP 3.4 x 10^{-7} mm Hg at 25°C. It has very low mammalian toxicity. It is recommended with 50 to 200 g ai/ha for control of variety of insect pests.

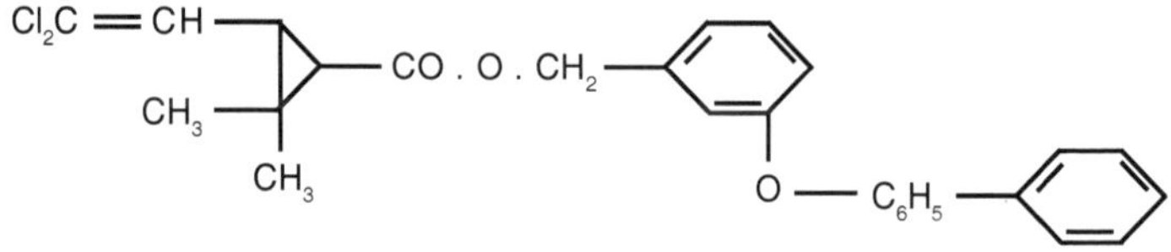

Permethrin

Cypermethrin ($C_{22}H_{19}C_{12}NO_3$)

It is discovered by Elliott and released by SIC comp with trade names "Ripcord", "Cymbush", "Imperator". This stomach and contact insecticide is chemically(RS)-a-cyno-3-phenoxybenzyl-(1RS)-cis,trans-3-(2,2 dichlorovinyl)-2,2-dimethyl cyclopropane carboxylate. Its technical form is visous, yellowish brown and soluble in acetone, ethanol, xylene, chloroform and water (only 0.01 - 0.2 mg/l). It is effective against a large number of insect pests with dose from 50 to 150 gram ai/ha.

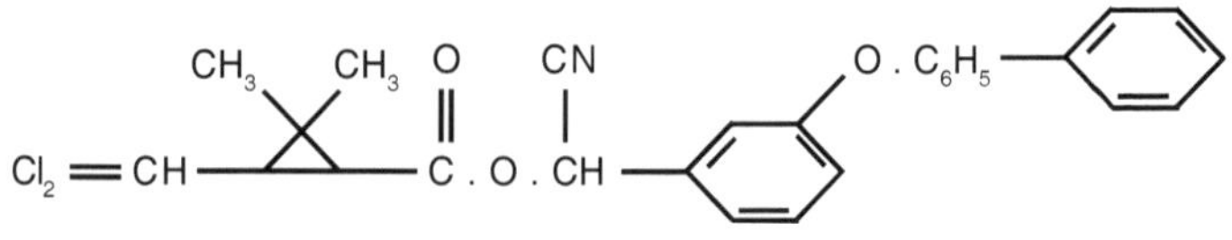

Cypermethrin

Fenvalerate ($C_{25}H_{22}ClNO_3$)

It was introduced in 1972 by Sumitomo Chem. Co. under the trade name "Belmark", "Sumicidin", etc. Its technical form is yellow viscous liquid with VP 2.3 x 10^{-7} mm/Hg at 20°C. It is slightly soluble in water and soluble in organic solvents. It is effective against a very large number of insect pests with dose ranging from 50 to 200 g ai/ha.

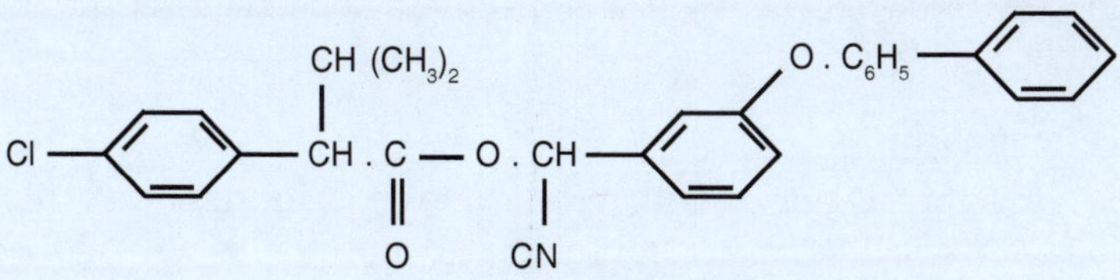

Fenuvalerate

Decamethrin ($C_{22}H_{19}Br_2NO_3$)

Its technical form is odourless, crystalline white powder with M.W. 505.2 and MP 98-101°C. It is insoluble in water. Comparatively it is more effective than other pesticides. It controls variety of insect pests with usual dose range 7.5 to 12.5 g ai/ha.

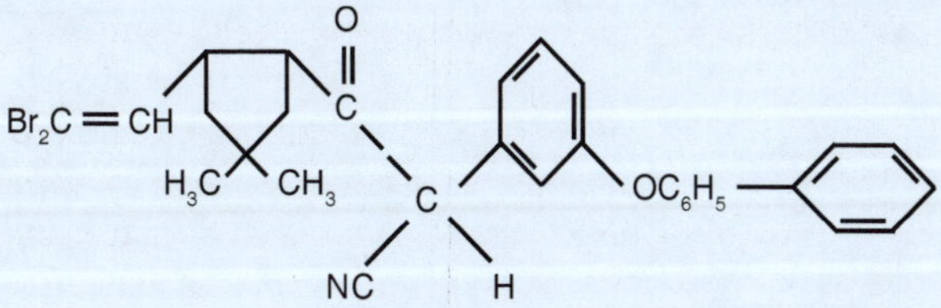

Decamethrin

Allethrin ($C_{19}H_{26}O_3$)

Chemically it is (RS)-3-allyl-2-methyl-4-oxocyclopent-2-enyl(IRS)-cis, trans-chysanthemate. It is sparingly soluble in water and soluble in most organic solvents. It is effective against household pests and other insects.

Bioresmethrin

Chemically it is 5-benzyl-3-furylmethyl(IR)-trans-chrysanthemate(1). This contact insecticide can control household pests and phytophagus pests. It controls mosquitoes, houseflies and cockroaches without or with mixing synergists (piperonyl butoxide).

Phenothrin

It is stomach and contact insecticide miscible in most organic solvents and water (2 mg/l). It is effective against the insect pests of public health.

Tetramethrin

It is contact poison with technical form, colourless, crystalline solid. It is effective against houseflies and mosquitoes and sandflies.

Resmethrin

All isomers of this compound are insoluble in water and soluble in solvents.

Bioallethrin

This is non-systemic insecticide.

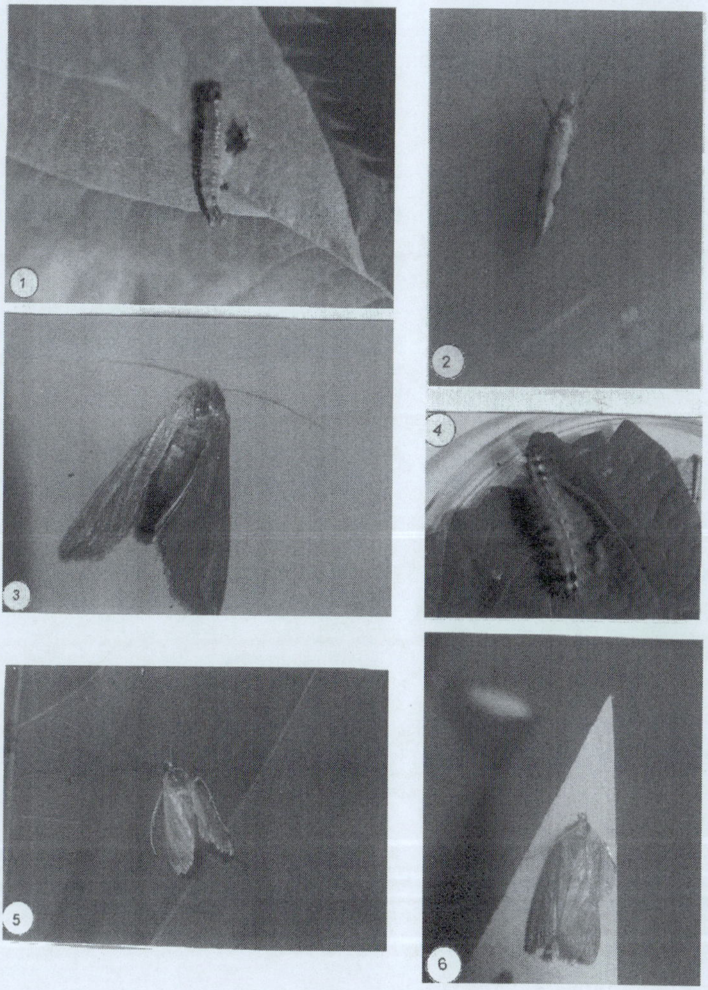

Fig. 1: **Cotton Leaf Roller *Sylepta derogata*.**

Fig. 2: **Diamond Back Moth *Plutella xylostella.***

Fig. 3: **Army Worm Moth *Mythimna separata.***

Fig. 4: **Bihar Hairy Caterpillar *Spilosoma obliqua***

Fig. 5: **Brinjal Fruit Borer.**

Fig. 6: **Maize Stem Borer *Chilo* sp.**

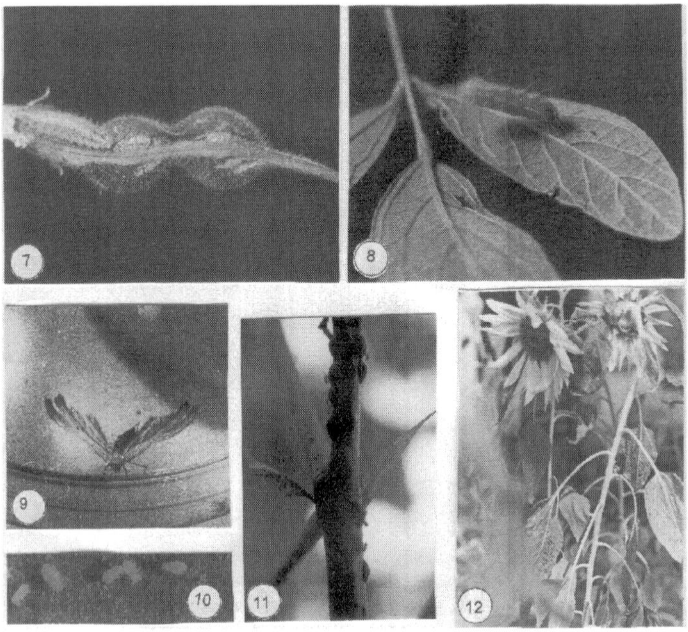

Fig. 7: Plume Caterpillar *Exelastis atomosa.*

Fig. 8: Plume Pupa *Exelastis atomosa.*

Fig. 9: Plume Moth *Exelastis atomosa.*

Fig. 10: Aphids.

Fig. 11: Cotton Scales.

Fig. 12: *S. obliqua* Larvae on Sunflower Leaves.

Chapter 2

CHEMOSTERILANTS

The control of insect pests is the most vexed problem of the day. Insects are the most formidable enemies of man since they destroy sizeable parts of crops, stored grains, household goods and live stock. Many insects are jeopardizing our health and hygiene.

Varieties of devices have been employed to control insect population, but none ever succeeded in exterminating them. Pesticides control the pests upto certain extent but create many serious problems. In fact, pesticides are creating more problems than that they solved. Hence, non-pesticidial control measures are reliable sources in the pest management.

The concept of decreasing populations of rapidly reproducing insects by reducing their birth rates rather than by increasing their death rates would worth in pest management strategies. Knipling, as early as 1937, for the first time postulated a method, the sterile male technique for insect control. He first investigated male sterilizing effect in the crew worm fly. *Cochliomyia hominivorax* in 1937 while working on the biology of the same insect. After second world war it was noted that in the genetics laboratory, certain insects were sterilized accidentally. However, Auerback and Robson (1942) first reported the chemicals interfering the insect reproduction. They reported that the fruitfly, *Drosophilla* could be sterilized by mustard gas (or nitrogen gas). This report has stimulated many workers for research on sterility of insects. Hence many reports were published from different parts of the world. Goldsmith *et al.* (1948) and Goldsmith & Frank (1954) reported that chemicals could retard ovarian development in *Drosophilla*.

In 1949 Bhattacharya reported sterility in the male *Drosophilla* by providing food contaminated with ethylene glycol. Bushland and Hopkins (1951, 1953) applied male sterile technique in the laboratory to control *C. hominivorax* flies. Later, Mitlin *et al.* (1954) recorded inhibitory effects of chemicals on the ovarian development of housefly. Similarly, Asher (1957, 1964) also demonstrated inhibitory effects in mosquitoes and houseflies. Since then the sterility principle of population control attracted the attention of entomologist at global level.

Definition

"Chemosterilants are chemicals which inhibit the growth of gonads or interfere with reproductive capacity of an insect".

Now a days, hundreds of chemosterilants are known and structural types are being discovered every year. The workers like Chang *et al.* (1966, 1967, 1970), Demilo and Borkovec (1968), Chang and Borkovec (1969), etc. contributed on the role of chemosterilants in metabolic fate of houseflies. The chemosterilants have reviewed with respect to new compounds, possible uses, biochemical characteristics, etc by the workers like Smith *et al.* (1964), Borkovec (1966, 1969, 1971), Campion (1972), etc.

Classification of Chemosterilants

The chemosterilants are classified variously by various workers.

Campion (1972) classified the chemosterilants in following way:

CHEMOSTERILANTS

1. Antimetabolites	5. Organometals
2. Biological alkylating agents	6. Insecticides
3. Aziridine derivatives	7. Insect hormone analogues
4. Phosphoramides and S. triazines	8. Miscellaneous compounds

Shrivastava (1996) classified chemosterilants as follows:

CHEMOSTERILANTS

Alkylating agents	Antimetabolites	Miscellaneous compounds	Plant products

Chemosterilants can also be classified as:

CHEMOSTERILANTS

Alkylating (Azaridines)	Non-alkylating	Dithiobiurets and Dithiazolium salts	Miscellaneous compounds
1. Apholate	1. Phosphoramides		
2. Tepa	2. Melamines		
3. Metepa			
4. Thiotepa			

Alkylating Agents

Labrecque (1961) investigated these compounds in USA. Alkylating agents are the most effective chromosome breakers; dominant lethal mutations are induced in all sperms (sterility). Lachance (1967) opines that chromosome abnormalities are found in all developing zygotes fertilized by such sperms and affect embryogenesis at the initial stages. Mitotic rate is depressed and mitosis is stopped in second/third cleavage division, much before the formation of a blastoderm. Some compounds of alkylating group cause mortality in larvae instead of embryos. This has been experimentally proved by Landa & Matolin (1971). Biochemical interactions of chemosterilants have been studied by Kilgore & Painter (1964) and Chamberlain & Barrett (1968). They reported that chemosterilants inhibit the synthesis of nucleic acid (DNA) and certain cellular enzymes and by that affect cellular growth. In general, alkylating compounds act on the insects by inhibiting nucleic acid (DNA & RNA) synthesis, inhibiting gonadial development and producing mutagenic effects. Alkylating agents are scattered in several groups of aziridinyls. The followings are the important among them.

Aziridines or Ethylenimines

Apholate (2,2,4,4,6,6,-hexakis(1-aziridinyl)2,2,4,4,6,6,-hexahydro-1,3,5,2,4,6,-triazatriphos-phorine):

Apholate is highly toxic to insects and mammals. It causes damage to bone marrow and germinal tissue in mammals and easily absorbed through skin. It is crystalline substance of white colour which is partially soluble in solvents like acetone, alcohol and chloroform. It remains stable in cool and dry conditions. Apholate is effective against houseflies at 0.5% when applied through food;

both, males and females are sterilized in housefly. It causes sterility in insects through alkylation. Apholate poorly inactivate the sperms in a parasitic wasp *Bracon* which is bio-control agent of agricultural pests. In newly emerged stable flies apholate treatment results in interference in the incorporation of Tritiated Thymidine into the DNA of the nuclei of nurse cells and follicular cells. Apholate is effective against housefly, mosquitoes, stable fly, screw worm fly, mexicanbean beetle, german cockroach and citrus red mite.

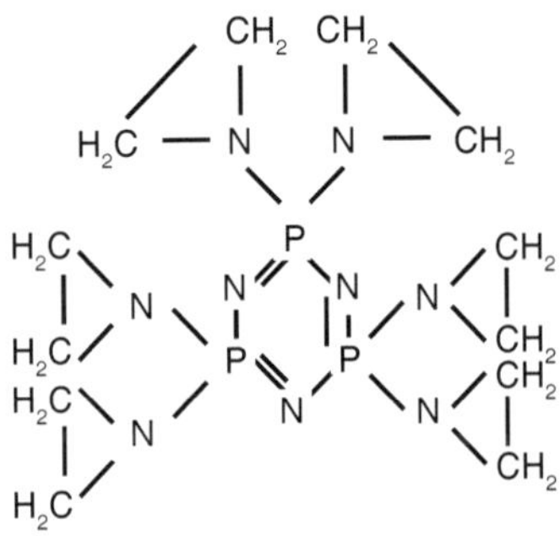

Apholate

Aphamide (N, N'-ethylenebis, [p,p-bis(1-aziridinyl) N''-methylphosphinic amide)

It causes damage to germarium condensation and pycnosis of nuclei, vaculation of cytoplasm and atropy of the follicular

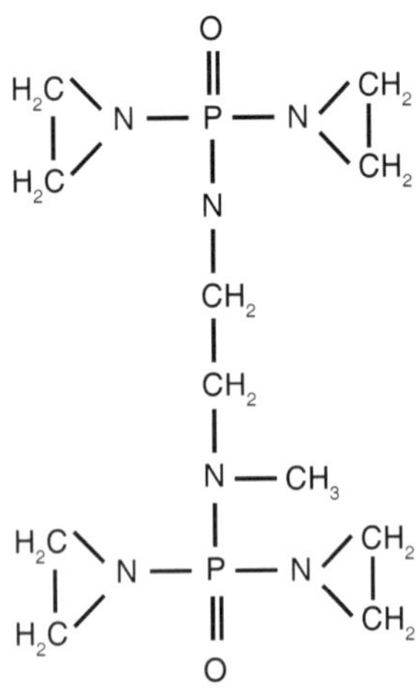

Aphamide

epithelium in *Musca domestica* (housefly). It also causes complete degeneration of the egg chamber in *Drosophilla*. In *Culex pipiens* extensive damage to germanaria and reduction in ovary size have been reported due to apholate treatment. *Aedes aegypti* showed the ultimate degeneration of gonadal cells and egg follicles. It is effective against housefly, mosquitoes, mexican fruit fly, german cockroach and citrus red mite.

TEPA (Tris(1-aziridinyl) phosphine oxide)

TEPA is also called as aphoxide. It is highly toxic to man. According to La Brecaue (1961) TEPA causes death to cancer patients when applied at 1.6 mg/kg. Chemically it is crystalline substance, which is extremely soluble in water. TEPA is odourless and colourless by nature. It is unstable and hygroscopic but highly soluble in acetone, ether and alcohol. It is effective against housefly, mosquitoes (Fig. 13), mexican fruit fly, german cockroach, screw worm fly, stable fly and citrus red mite. It inhibits gonadal development in housefly. It also acts as sperm inactivator in insects. It sterilizes both the sexes of housefly when applied 0.1 per cent TEPA through food. TEPA also produces sperms inactivation in citrus red mite and acts as useful source of control of red mite on citrus crop. In cockroach, treatment of TEPA results in reduction in the number of basal oocytes and gradual degeneration of the ovaries. In certain beetles degenerative effects of TEPA have been observed.

As like apholate, TEPA damages germarium condensation and induces pyc nosis of nuclei in housefly. Vacuolization of cytoplasm and atrophy of the follicular epithelium is also noticed in *M. domestica*. It can sterilize mexican fruit fly with 5% tepa solution in

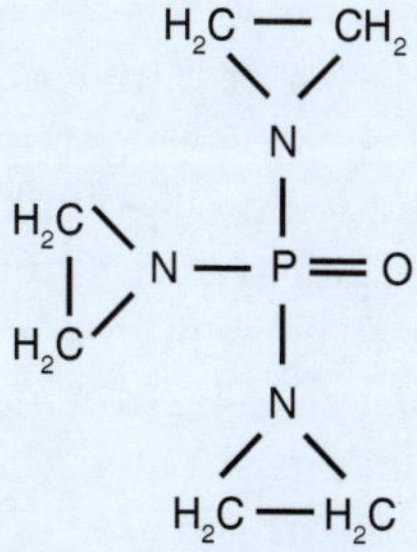

TEPA

methanol. TEPA is also responsible for reducing the rate of malaria transmission when mosquitoes infected for more than 14 days were held on a TEPA residue of 10 mg/sq ft. TEPA sterilizes both sexes of a teak defoliator, *Pyrausta machaeralis* (Walk) (Lepidoptera).

METEPA

It is straw coloured liquid. It dissolves completely in solvents and water. It is easily absorbed through the skin of mammals and insects. It is more stable in storage. It causes sterility in both the sexes of housefly and a teak pest, *P. machearalis*. It is also effective against stablefly and screw worm fly. The degenerative effects of metepa on the telotrophic ovaries of certain beetles have been demostrated by Novak *et al* (1968).

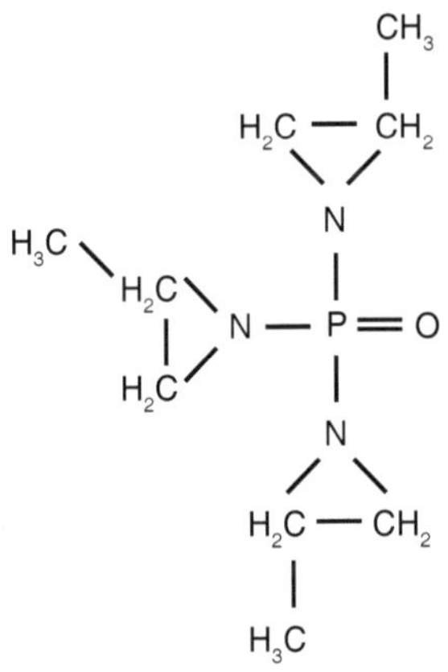

METEPA

Thio-TEPA

It is sulphur analog of tepa. It induces mutagenic changes in oocytes. Thus, acts as good chemosterilant in the insects. It is white crystalline solid and comparatively less soluble in water and solvents than metepa. It is effective against house fly, cockroach and screw wormfly. Thiotepa reduces the size of ovary significantly in *Dacus dorsalis* (Fruit fly). Thiotepa causes contraction, vacuolization and degeneration of ooplasm leading to arrest of yolk formation and subsequently immaturation of oocytes. Pycnosis in follicular epithelial cell nuclei and nurse cell nuclei, shrinkage in follicular epithelial layer, degeneration of germinal regions and fragmentation in nurse cell nuclei are the prominent deformities in *D. dorsalis*. Thiotepa causes the damage to germinarium condensation and induces pycnosis of nuclei and also shows cytoplasm vaculation and atrophy of the follicular epithelium in *M. domestica*. Recently, it is proved that thiotepa is excellent source for sterilizing males of *Culex fatigen* mosquitoes.

Phosphine oxide and Tris (2 methyl 1-aziridinyl) Tepa are also important insect chemosterilants grouped under alkylating compounds.

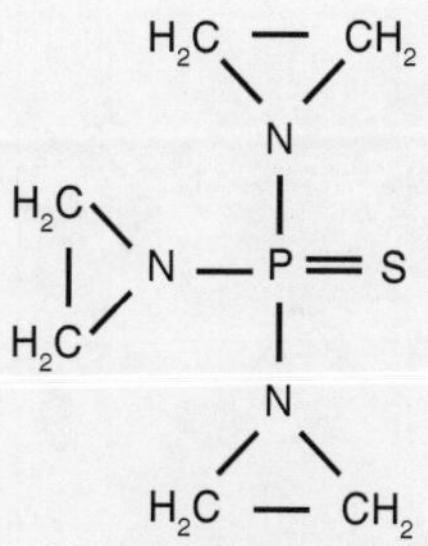

Thio--TEPA

Non Alkylating Chemosterilants

Alkylating chemosterilants comparatively are not species specific hence, non-alkylating chemosterilants came forth in an attempt to find safer chemicals in pest management. Non-alkylating chemosterilants preferably attack rapidly dividing cellular systems. These in adult insects are found in reproductive system. Such systems serves as targets for chemosterilant interactions. Non-alkylating chemosterilants are visualized by following four groups namely:

1. **Phosphoramides**
2. **Melamines**
3. **Dithiobiurets**
4. **Dithiazolium salts**
5. **Miscellaneous compounds.**

Phosphoramides

Hempa (Hexamethyl phosphoramide)

It is an important chemosterilant which causes sterility in housefly males. According to Morgan (1967) Hempa is responsible for destruction of egg chamber in *M. domestica*. Physiologically it interacts with the insect as like in alkylating chemosterilants. According to Chang & Borkovec (1969) Hempa and Tepa shows equal mortalities in housefly. However, Tepa is 50 times more effective than Hempa in causing sterility. Hempa sterilizes both sexes of *P. machearalis* and *Periplanata americana* (Fig. 14).

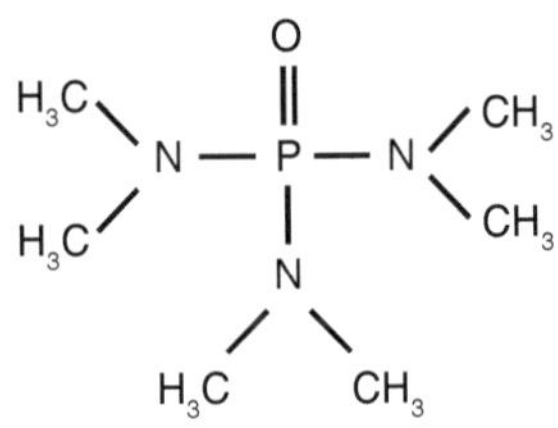

Hempa

Thiohempa

Thiohempa is sulfur analogue of hempa. It is also an effective chemosterilant. It sterilizes sugarcane stalk borer with 2.5% and 5% pupal dip.

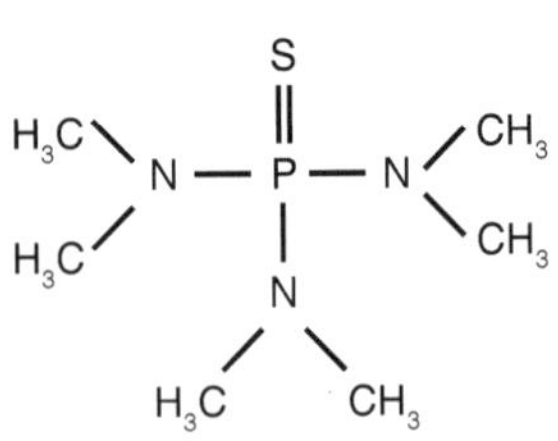

Thiohempa

Hemel (Hexamethyl emelamine)

It is an important chemosterilant of male housefly. Its chemical structure is shown below:

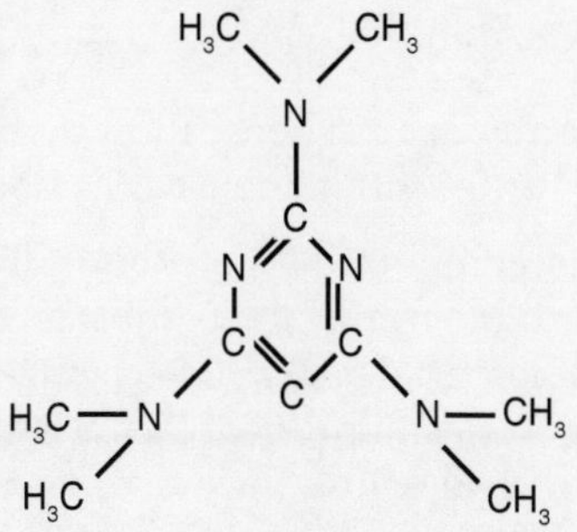

Hemel

Melamines

Chemically these are Thiamino-1,3,5-triazines. Due to activation they are very effective chemosterilants for houseflies.

Dithiobiurets and Dithiazolium Salts

A dimethylamino compound 3,5-bis (dimethylamino) 1,2,4-dithiazolium chloride was first developed by the Hreculus Powder Company. There are several analogues available in the market now.

Antimetabolites

The antimetabolites from pyrines and pyrimidines include:

1. Uracil
2. Uracil, 5-bromo-
3. Uracil, 4,5-diamino-, sulfate
4. Uracil, 5-fluoro-

All above compounds are effective against housefly. However, uracil, 5-fluoro- is more important since it affect oriental fruit fly (*Dacus dorsalis*), mexican fruit fly and screwworm fly.

The antimetabolites other than pyrines and pyrimidines include:

1. Methionine
2. Thymine
3. Tyrosine
4. Methotrexate, and

5. Glutamic acid, N- {p-{ [(2,4-diamino-6-pteridinyl) methyl] amino} benzoyl}.

Methotrexate is effective against house fly and screwworm fly while, rest of the members are effective against only house fly.

Antimetabolite compounds are chemically and structurally related or similar to biologically active substances. Their structural similarity confuses biological system to recognize them from natural substances hence not used at proper place. They interfere with the synthesis of the nucleic acids. Purine compounds, pyrimidine compounds and folic acid compounds are the important groups of antimetabolites which play an important role in affecting population of insects. These are tabulated in Table 1.

Table 1: Antimetabolities for Insect Pests

Sr.No.	*Purine Analogues*	*Pyrimidine Analogues*	*Folic Acid Analogues*
1.	5-Bromouracil	2-amino-4-methyl pyrimidine	Aminopterin
2.	5-Fluorouracil (5-FU)	1,2,3,6-tetrahydro-2,6-dioxo-4-pyrimidine acetic acid	Amithopterin
3.	5-Fluoroorotic acid (FO)		Methotrexate
4.	2-Thiouracil		Folic acid
5.	2,6-Diaminopyrine		

Antimetabolites interfere with the synthesis of the nucleic acids. They compete with the normal nucleotides for enzymes and prevent sequence of the nucleotides into nucleic acids. They exert a feed back control on the synthesis of the nucleotides themselves. Antimetabolites directly incorporated into the nucleic acids and interfere with synthesis of nucleic acids. 5-fluorouracil replace the normal uracil from RNA of eggs when fed with insect diet. This affect the viability of eggs and further, population. Thus, antimetabolites functions as analogues of purine and pyrimidine bases occurring in nucleic acids.

Miscellaneous Compounds

These compounds are structurally unrelated to each other. Some chemicals like folic acid, antagonists and mitotic poisons have good sterilizing effect on insect pests.

Aminopterin

Aminopterin is an antagonist which cause sterility in insect pests. It causes sterility in house fly female at 0.01 per cent concentration but, can not induce sterility in males. It does not cause sterility in screwworm fly males. Houseflies fed with aminopterin at the rate of 0.2 mg/ml in 50 per cent skimmed-milk-aqueous solution, exhibited a marked reduction in ovarian growth.

Mustard gas

It acts as mitotic poison and is effective against fruit flies.

Organotin compounds

Many triphenyltin derivates are reproductive inhibitors for houseflies. These compounds cause sterility in both sexes of house fly but, males require higher concentration dose. Important examples are give below:

Triphenyltin chloride,

Triphenyltin acetate,

Triphenyltin hydroxide, etc.

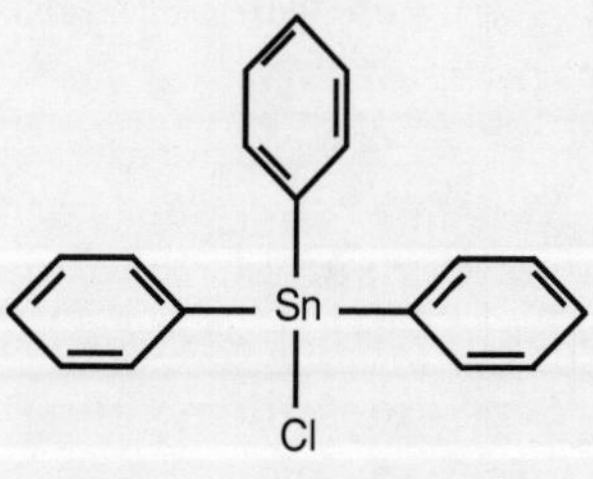

Triphenyltin chloride

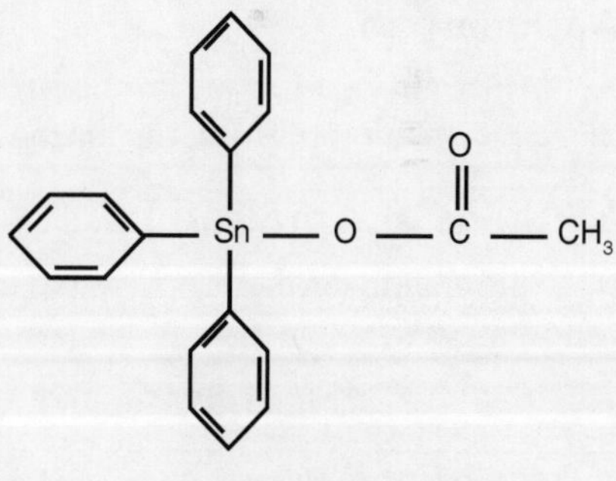

Triphenyltin acetate

Pesticides

Aldrin: It is an effective chemosterilant against house fly.

Acetamide, (2-naphthyloxy)

It is effective against house fly.

Acetic acid, chloro, 6-pentyl-m-tolyl ester

It is effective against houseflies.

Benzene boronic acid

It is effective against houseflies.

Plant products

Colchicine: It is effective against housefly and screw worm fly. It shows antimitotic effects in insects.

Coumarin: It is effective against houseflies.

Monocrotoline: It shows antimitotic effect against insects.

Essential oils: Extracted from plants like coriender, clove, eucalyptus, cassia, acorus, etc. EO acts as ovicide and affect insect reproduction.

Pyperonyl butoxide

It is effective against houseflies.

Urea, 2-thio

It is effective against houseflies.

Anthramycin

Anthramycin is a antibiotic drug. It sterilizes houseflies and drosophylla by inhibiting biosynthesis of nucleic acids.

Methods of Application

The chemosterilants applied to the insect pests are classified in following way:

CHEMOSTERILANT APPLICATIONS

Dusts and Sprays	Insecticide Mixtures	Direct Contact Residues	Ora ingestion

Criteria for Effective Applications (Methods)

The effectiveness of the method of application of chemosterilants is dependent on the following aspects:

1. Males of the insect pest species must be very mobile.
2. When sterilized the insect must not lose its sexual urge of mate, and

3. The female must be satisfied with the act of mating rather than fertility.

Methods

As dusts and sprays

With the help of sprayers and dusters chemosterilants are treated to the objects like ponds, lakes, reservoirs, open ditches and other essential places.

As insecticide mixtures

The chemosterilants are mixed with insecticides in appropriate proportion and concentration for their use.

As contact residues

The chemosterilants residues are applied directly to the places where insects like to rest and visit frequently for food, shelter and mate.

Oral ingestion

The chemosterilants are given to the insects through their natural or artificial food in laboratory or in field condition.

Advantages and Disadvantages

Advantages

1. Helps in minimizing the use of conventional pesticides which cause pollution and pest resistance, secondary pest-out break, pest resurgence, etc.
2. Cheaper than irradiational method.
3. Offers higher mortalities in pests than conventional pesticides.

Disadvantages

1. Dangerous to higher animals, as they produce mutagenic effects and sterility.
2. Carcinogenetic to man.
3. Phytotoxic.

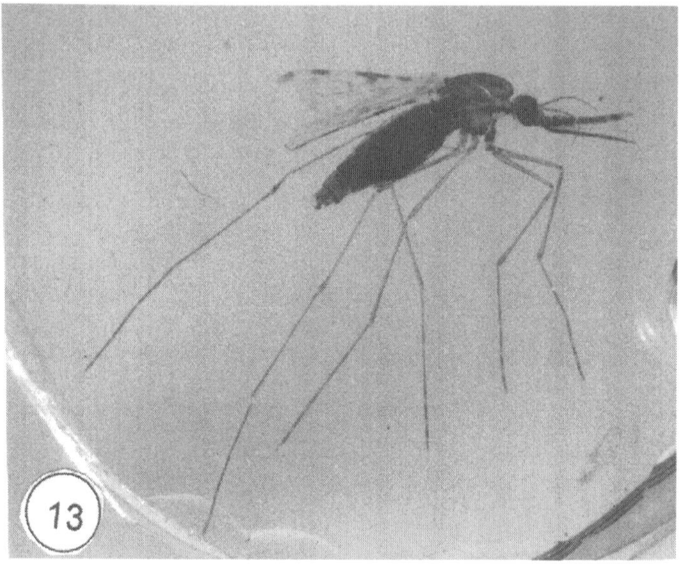

Fig. 13: Mosquito – *Aedes* sp.

Fig. 14: Cockroach *Periplanata americana.*

Chapter 3
RADIATIONAL PEST CONTROL

Radiational pest control is the suppression of target species below the level of economic damage by sterilizing them with different radiants. Radiational control is useful against pest insects, microorganisms and quarantine. Within recent years, advances in our knowledge have opened an entirely new vista for enduring research work in solving pest control problems. Knipling, as early as 1937, for the first time demonstrated a new method, sterile male technique for the control of insect pests. If an overwhelming number of sexually sterile males would find out the native females, and convert them reproductively ineffective by mating. Such repeated release would soon achieve control or eradication of the pest species. In the laboratory, Bushland & Hopkins (1951) first experimented with the screw worm fly, *Cochliomyia hominivorax* (Coquerel). Later Baumhover *et al.* (1955) and Lindquist (1955) practiced these experiments in a small area of the field of Island (Curacab). They successfully eliminated screw worm population from the region and thus showed the potential of the application of sterile male technique in pest management. However, Knipling (1960) and Lindquist (1961) got outstanding success in eradication of screw worm flies from a very large area of Florida, USA for which Knipling received price from president of America. Since then radiational control attracted the attention of Entomologists for supression of notorious insect pest population. However, up to day, no such outstanding success have been achieved so far through radiational control in the world.

Hence, there is need to undertake more research on radiational control.

The review of literature indicates that the workers such as Knipling (1937, 1955, 1960), Procter *et al.* (1954), Kraybill (1959), Dennis (1961), O'Brien & Wolfe (1964), Cornwell *et al.* (1966), Titoon *et al.* (1966), Cogburn *et al.* (1966, 1972), Proverbs (1969), Borkovec (1971) and Shreenivasan *et al.* (1972), etc have contributed on radiational pest control in insects. Cornwell *et al.* (1966) studied merits and demerits of radiational control and radiational effects on the pest species. Titoon *et al.* (1966) used gamma radiations against several stored gain pests. Likely Coghurn *et al.* (1966) also worked on the same insect group and the radiational source. Dennis (1961) studied dose dependent mortality in several orders of class Insecta. For lepidopterous insects minimum dose of 40 krad is required to induce sterility. In order Diptera it is just 3-5 krad while in Coleoptera induced sterility is seen with radiational dose of 6-16 krad. O'Brien & Wolfe (1964) studied the radiation impact on insect pests. They reported that radiation impact is dependent on age, sex and nutritional status of the pest species. Temperature and humidity are also important factors for influencing effectiveness of radiation. Coghurn *et al.* (1972) and Shreenivasan *et al.* (1972) studied control aspect of stored grain pests through radiational effect. But it has come to know that radiational control is not fisible in stored grains since it has several demerits such as lack of facilities, other control methods easier than radiational, reinfestation is still possible and it is costly than other control measures, etc.

In India, BARC is actively engaged in doing research on radiational control of insect pests. From BARC, Rahalkar & Nair (1968) and Rahalkar & Lewis (1969) worked on several insect pests of stored grains for their control by radiational methods. Shreenivasan *et al.* (1972) also added significant contribution on radiational control of stored grain pests in India.

Methods

Basically there are two methods of radiational control.

1. Direct sterilization method,
2. Indirect sterilization method.

Radiational control is designed by several radiant (solar) energy sources. Important among them are given below:

1. X-rays and γ-rays,
2. Ultraviolet and visible rays,
3. Infrared light, and
4. Radio frequencies.

X-rays and γ-rays

These sources produce chemical effects including ionisation of atoms of the absorbing media. These are effective against stored grain pests. X-rays and γ-rays interact with insects by directly killing them or sterilizing them to supress their population in future generations below the level of economic damage. γ-rays produces Co^{60} source which is more penetrative and promising to control stored grain pests. Economically Co^{60} radiational method is costly and hence it is less profitable.

Male Sterile Technique

The experimental insects are first reared in the laboratory. Then they are sterilized by exposing to gamma radiation (Co^{60}) and then released in the field with greater number than that existing in nature. The laboratory sterilized males are more competent than wild and get more chances to mate with females and render them reproductively ineffective.

Important Characters of Sterile Insect

Ideal sterile insect should have following features:

1. It should be sexually vigorous.
2. It should be long lived.
3. It should be adaptable to field conditions.

In order to achieve full success by this method, a good knowledge, however is needed on the biology, population dynamics, dispersal, ecology, flight range and behaviour of the insect to be controlled.

The sterile male technique is illustrated in the following hypothetical but realistic model proposed by Knipling (1964, 1968).

Table 2: Relative Merits of the Two System of Control

Generation	Uncontrolled natural population (5 × increase rate of each sex)	Controlled natural population each sex	Sterile Insect Technique			Controlled Insecticides treated (90% kill)
			Sterile male released	Ratio of sterile to fertile	Insects reproducing	
1.	1,000,000	1,000,000	9,000,000	9:1	100,000	1,000,000
2.	5,000,000	500,000	9,000,000	18:1	26316	500,000
3.	25,000,000	131,625	9,000,000	68:1	1,907	250,000
4.	1,25,000,000	9,535	9,000,000	942:1	10	125,000
5.	6,25,000,000	50	9,000,000	18,000:1	0	62,500
6.	6,25,000,000	0	-	-	-	34,250

The above model is assumption based. In a specific area, suppose the natural population of a species to be eradicated is 1,000,000 of each sex and it shows five times increase at each generation, and getting stationary after some time due to biotic factors. Uncontrolled population of the fifth generation would increase to a maximum of 6,25,000,000. But at each generation if 9,000,000 fully competitive sterile male insects are released, in the beginning there would be a dominance of sterile insects in the population in the ratio of 9:1, sterile to fertile. Due to competitiveness only 1/10 of the total insects (100,000) are expected to be reproducing and giving rise to 500,000 individuals in the next generation. Additional released of new batch of 900,000 sterile male would result in about 26,316 capable individuals of normal reproduction. Thus as at successive generations there would be considerable decrease in the natural population and in the fifth generation the population would be completely eliminated. While through insecticide treatment 90 per cent killing of the pests will be achieved slowly and will require 18 generations. Thus sterile male technique is far better than conventional pesticides (Table 2).

Sterilization by irradiation is more effective and rapid method of pest control. However, it has its own limitations. It has high cost involved in the equipments usually the somatic cells exhibit greater, susceptibility to radiation than the reproductive organs. Hence, few species of insects have been successfully sterilized by radiation without losing appreciable competitiveness (Borkovec, 1971). Several workers from different parts of the world are actively engaged in doing research on radiational control of insect pests. The list of insects and the countries involved in the radiational control is given below:

Table 3: Work on Radiational Control in Different Countries

Sr.No.	*Insect Species*	*Country*
1.	*Tribolium* spp.	India
2.	*Sitophilus granarius*	Iran
3.	*Corcyra cephalonica* (Rice moth)	Thailand
4.	*Sitotroga cerealella*	Paraguay
5.	Mosquitoes, *Culex* spp. & *Anopheles* spp.	India
6.	Tsetse fly, *Glossinia* spp.	South Africa
7.	Fruit fly, *Dacus* spp.	Several European and Non-European Countries, Latin America, South-East Asia etc.
8.	*Rhyncophorus ferrugineus* (Palm weevil)	India (BARC)
9.	*Zabrotis subfaciatus*	Brazil
10.	*Heliothis* complex (Fig. 17)	Latin America, India
11.	Boll worm (Fig. 16)	India & other countries
12.	*Cadra cautella*	Iraq, Iran etc.
13.	Maize borers (Fig. 15)	India, Indonesia
14.	*Scirpophaga nivella* (Sugarcane borer)	Indonesia
15.	Codling moth, *Dacus dorsalis, D. cucurbitae*	India

Ultraviolet (UV) and Visible Light

UV to blue green (300-560 nmλ) region of the solar spectrum is most attractive for insects but red orange and yellow are not attractive. Yellow colour is repulsive for most of the insects hence it is used to repel insects and UV lamp is used to attract nocturnal insects. Man can see the colour violet to red (380-770 nmλ) of the solar spectrum but insects have range between UV to red (253-700 nmλ).

Visible light can be utilized in pest management in following ways:

1. By producing phototaxis,
2. By inducing diapause by altering the photoperiod, and
3. By modifying ovipositional behaviour.

Phototaxis may be positive to attract the insects or negative to repel the insects. Many economic pest species get attracted to light and are punished to death by trapping them with various types of traps. Corn borer, tobacco borer, tomato horn borer, citrus moth, etc. are effectively controlled with light traps in USA. The onset of insect diapause is prevented by extending the length of the day by flooding the light. Codling moth of apple crop is controlled by this way. Control of codling moth of apple crop by this method is usual practice in USA wherein, apple plants are exposed to extra artificial light which interferes with the egg laying of the moths and this significantly reduces the damage to apple by codling caterpillars.

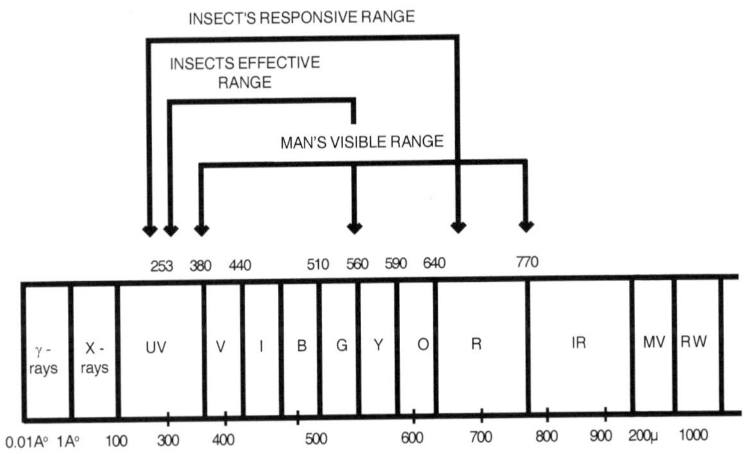

Visible ranges of insects and man in the solar spectrum.
N.B. Wavelength are in nanometer (nm) except where mentioned.
1 µm = 1/1000 mm
1 nm = 1/1000 µm
1 A° = 1/10,000 nm.

Infrared (IR)

Infrared (IR) is used for direct killing of the insects or for attracting the insects. IR radiation is intermediate between RF and a visible part of the electromagnetic spectrum (wavelength 10,000 A° or 1 μ). IR radiation is produced from hot bodies like IR lamps. Certain pests like lepidopterous moths and dipterous mosquitoes possess IR sensors in their antennae. Hence, they are attracted to an IR source and then killed by conventional methods.

Radio Frequencies (RF)

Radio frequencies produce heat and cause mortalities in the insect pests. It is of longer wavelength. RF is good source for controlling stored grain pests and wood infesting beetles like *Lyctus* sp. It produces differential healing in stored grain pests since there is difference in the moisture content, humidity and dielectric properties of the two. The RF healing cause the death of the insect pest by synaptic blocking, vacuolisation, coagulation of nervous system or change in amino acid content. However, it does not cause much damage to the seeds. RF method of pest control is more effective against stored grain pests like *Sitophilus* spp., *Tribolium* spp., and *Trogoderma* spp.

Culex pipens fatigans is radiosterilized (Sharma, 1981). The dose of chronic Co^{60} gamma radiation decreased egg production in a parasitic wasp *Bracon hebetor* Say. The effect of X-ray dose with component of mixture cause most destruction to egg in above parasitoid. Lowered egg production is also noticed in holometabolic insects due to irradiation. The vulnerability of developing ovarian tissue has also been demonstrated for hemimetabolous insects like grasshoppers with respect to fast neutrons, gamma rays and X-rays (Grosch, 1962). A comparative study of X-rays, β-rays & γ-rays demonstrated that X-rays were most effective in reducing the percentage of egg hatching in *Bracon hebetor.* Alpha radiation has also been used on newly laid *Bracon* eggs. It has been observed that the egg nucleus is inactivated by one alpha particle in contrast to only 50 per cent lethality when 16×10^6 particles pass into the cytoplasm.

Ionising radiations are recognized to induce dominant lethal effect in gamates. In *Drosophilla* dominant lethals appear to be mixture of events "not all involving breakage and/or loss of genetic material". Same situation is noticed in *Bracon hebetor* also. Radation induced delay in development of many insects. In *Habrobracon,* extensive abnormality in females X-rayed just before pupation, exhibited one deformed antenna, all six legs misshapen, a small medial ocellus and an abnormal stinger. Developmental abnormalities are also noted in *Drosophilla* and *Bracon* when induced by ultraviolet radiation and X-radiation resepectively.

Radiational mutational reports are published on mosquitoes and *Drosophilla.* A moderate number of hymenopteran mutational studies have also been reported (Whiting, 1934) as changing in size, morphology, pigmentation, mosaicism, behaviour and sex. Radiational mutatic effects are also reported in *Lymantria, Ephestia, Tribolium* etc and orthopteran pests.

A wood inhabiting beetle, *Lyctus planticollis,* LeConte, appeared to die two days after sizable doses of Co^{60} gamma rays. In chalcids (parasitoids) heat preconditioning increases tolerance to X-rays but conversely, a non-lethal dose of X-ray followed by a non-lethal heat shock results in death. Photomeric observations on X-rayed grasshopper embryos shows an increased dispersal of DNA in nuclei which become enlarged and may lead to mutatic changes.

Lindsay & Craig (1942) widely studied radioisotopes distribution in insects. The isotopes they fed p^{32} got concentrated in the midgut epithelium and gonads which showed mutagenic effects in certain insects. In *Bracon* p^{32} induced dominant lethal to males. P^{32} has also showed effects on mosquitoes species, Drosophilla and housefly. Developmental effect of ingested radiophosphorus have been demonstrated for mosquitoes, houseflies, pomace flies reared on radioactive media and for wasps reared on radioactive host caterpillars. Herbivorous insects accumulate, Cs^{137} to an extent which in ionic content come to resemble the plants of contaminated area, while Sr^{90} content is lowered. In *Romalea microptera* most of the Cs^{137} ingested from bean plants get concentrated in the muscular tissue rather than digestive tract and reproductive system thus affecting the physiology.

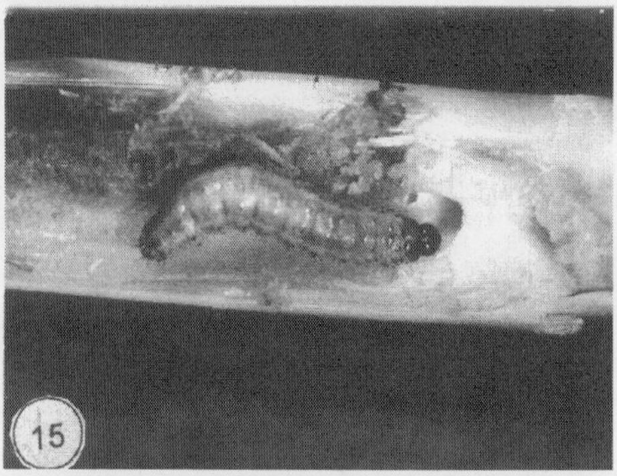

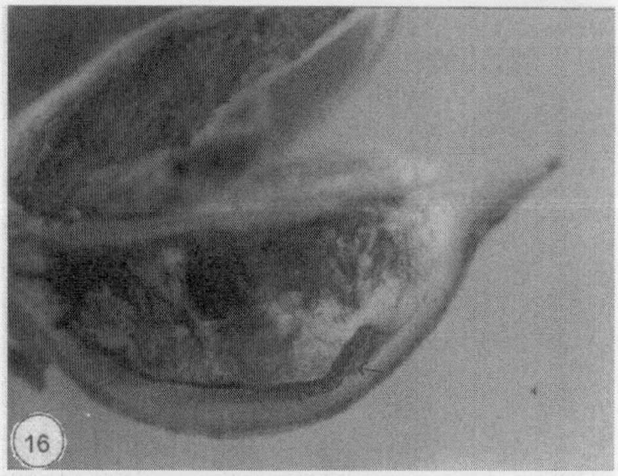

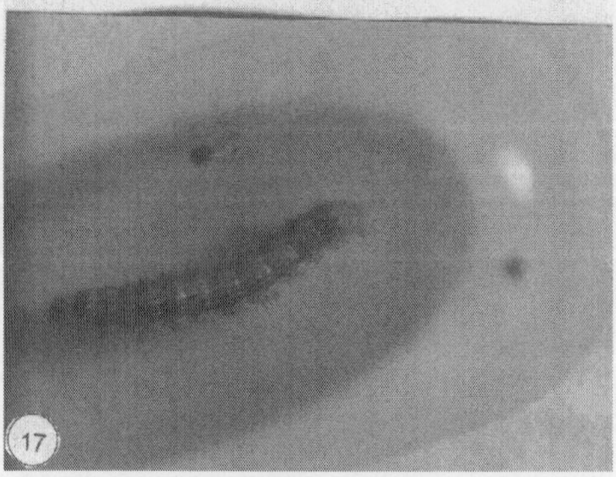

Fig. 15: Jowar Stem Borer *Chilo partellus*.

Fig. 16: Pink Boll Worm *Pectinophora gossypiella*.

Fig. 17: Gram Pod Borer *Helicoverpa armigera*.

Chapter 4

PHEROMONAL PEST CONTROL

It is now common knowledge that chemical pesticides for the control of insect pests are no longer the powerful weapons they once were. In fact, pesticides cause several serious problems like health hazards, pollution, pest resistance, secondary pest-out break, poisoning the entire food chain, and many more demerits. On this background, pheromonal pest control would prove its worth in pest management strategies.

Pheromones and chemical ecology play an important role in pest management. Chemical ecology is concern with communication of signals through specific chemicals between organisms in an ecosystem. Semiochemicals are signalling chemicals that an organism can detect in its environment which may affect the organism's behaviour or physiology. Those that act between members of the same species are called pheromones and those that act between species are called allelochemicals. The allelochemicals may be allomones, which favour the emitter and kairomones, which favour the receivers. Many allelochemicals act as both allomones and kairomones hence called "Synomones". However, insect sex hormones are the best known semiochemicals.

The practice of using sex pheromones as monitoring tool in integrated pest management (IPM) strategies is well recognised world wide. Sex pheromones released by insects serve as attractants for their conspecific males/females so that mating may be ensured between the same species (Tamki, 1985). The insects which depend upon pheromonal communication in their mating usually have a highly developed olfactory sense for them because the pheromones are released in a very minute quantity. Pheromones can be used for sampling of insect pest populations (survey studies) and behavioural manipulation. Both are important components of pest management.

The population survey through pheromonal component is helpful in pest management for understanding economic damage level and time of pest build up and further suggesting the appropriate control measures.

Behavioural responses of pests in the form of normal attraction due to pheromonal components towards reliser can be disrupted for causing death of the species. This objectives can be achieved by (i) use of pheromone-baited traps, (ii) orientation to a source of chemosterilant or insecticide, and (iii) orientation to a combination of light and pheromone.

Through introduction of synthetic pheromones into the environment it is possible to disrupt normal communication between the species and to prevent mating. Disruption can be made to sensory functions, central nervous system functions, habituation, and confusion. By parahormones and antihormones also natural pheromonal communication is disrupted. Parahormones are structurally similar to pheromones and may have same response to female from the target species (caproic acid parapheromone identified for honey bees). Antipheromones help in maintaining reproductive isolation between the closely related species. They disrupt mating and pheromonal communication and thus affect pest populations.

Definition

Pheromone is exocrine ectohormone in the form of a chemical or mixture of chemicals which released by an organism in environment and cause specific reaction in a receiving organism of the same species. Pheromones are chemical messengers for a species. The species may respond specifically to the source. Karlso & Butenandt (1959) gave the name to these substances as "Pheromones". (Gk, Pherein = to carry; hormon = to excite).

Classification

SEMIOCHEMICALS (Behaviour Regulating Chemicals)

- Pheromones
  - Primer Effect (Gustatory)
  - Releaser Effect (Olfactory)
  - Mono Component
  - Multi Component
- 1. Sex Pheromones
- 2. Aggregation Pheromones
- 3. Alarm Pheromones
- 4. Trial Pheromones

- Allelochemics
  - Allomones
  - Kairomones
  - Synomones

Allomones

Allomones are chemicals or mixture of chemicals that are released by one organism to induce a response in an individual or another species and is advantageous to releaser. *Example*: Ants and bees.

Kairomones

Kairomones are chemicals or mixture of chemicals that are produced by one species to induce response in an individual of different species is advantageous to the recipient. *Example*: Nezara bug – Tachinid fly.

Chemical communication between insects has attracted great interest in recent year and there is a sufficient literature available. The communication by means of pheromones is almost universal within the animal kingdom but it has reached its highest degree of development and complexity within the insect. Pheromones play a major role in integration of social insects and controlling insect reproduction.

Types of Pheromones

Pheromones are of following types:

1. Sex pheromones,

2. Aggregation pheromones,
3. Alarm pheromones and
4. Trail pheromones.

Sex Pheromones

Pheromone Producing Glands

Pheromone producing glands are located on several parts of the insect body such as head, thorax, legs, wings and abdomen depending on the insect pest species. These glands are ectodermal in origin and exocrine hence, they open and release their products outside the body. Thus, pheromones are said to be ectohormones directly released in the atmosphere as chemical messengers.

Sex pheromones are widely studied in several insect orders like Lepidoptera, Hemiptera, Coleoptera, Orthoptera, Diptera, Dictyoptera, Hymenoptera, Neuroptera and Mecoptera. However, Lepidoptera is on its evolutionary peak in sex pheromones. In male butterfly of *Danaus* spp. aphrodisiac gland is a source of sex pheromones while, in some lepidoptera glandular scales, "androconia" on the wings is good source of sex pheromones for males. Aphrodisiac gland is not only the source of sex pheromones in Lepidoptera but also in Diptera, Hymenoptera, Neuroptera and Trichoptera. Several other insects may also have the specialized glands for sex pheromones.

Mechanism of Pheromone Reception

There are three important components of pheromone reception in insects which includes:

1. Pheromone producing glands,
2. Pheromone travelling medium, and
3. Pheromone receptors in insects.

Pheromone producing glands

These are exocrine glands. Pheromone producing glands are already discussed in this chapter.

Pheromone travelling medium

Pheromones are carried from emission point to reception point through a medium of air or water.

Pheromone receptors

Pheromone receptors may be olfactory (smell) (Fig. 18b) or gustatory (taste) (Fig. 18a). Sense organs are present on insect body.

Olfactory receptors

ORS are of four types. The olfactory sensilla are mostly located on the antennae of the insect but, in some cases they are found on maxillary and labial palpi. Sexual diamorphism is always recorded in pheromone receptive insects specially with the structure of the antennae. The antennae of pheromone receiving male moths are larger and greatly branched than the females. Such specialized male antennae accommodate the numerous olfactory sansilla for better communication.

In the insects, mainly four types of olfactory sensilla have been identified those are listed below:

1. *Sensilla trichodea*: Found in Lepidoptera
2. *Sensilla basiconica*: Found in Diptera, Lepidoptera, Hymenoptera and Coeloptera.
3. *Sensilla placodea*: Always found in order Hymenoptera.
4. *Sensilla coeloconica*: It is found in orders, Hymenoptera and Orthoptera.

Trichodea sensilla (olfactory sensilla) (Fig. 18b)

The structure of trichoid sensillium is represented in the Figure 18b.

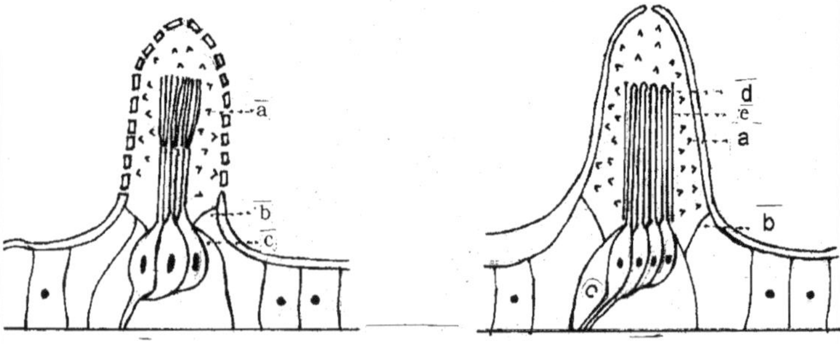

(a) (b)

Fig. 18 (a): Gustatory sensilla, (b) Olfactory Sensilla.

a – Receptor lymph, b – Outer sheath cell, c – Inner sheath cell, d – Dendrite, e – Dendrite sheath.

Trycoid types of sensillum is located on antenna of an insect. Basement membrane, epidermis and cuticle are important components of body wall involved in trycoid type of sensillum. Cuticle shows large number of pores ranging from 1000 to 50,000 per sensillum (silk worm). The pore number is vary with the species. The pores open into lumen of the sensillum. Lumen contains a fluid called sensillum liquor which bathes the dendrites. The liquor absorbs the pheromone molecules. These molecules then move to acceptor (receptor) site on the dendrites due to diffusion and initiate transduction. Transduction is the process in which transformation of the stimuli is made into electrical impulses. The impulse later reach to the brain. Insect olfactory system is well developed and more sensitive than man made electric detection system. Hence, still we know very little about the above process.

Chemistry of Sex Pheromones

Sex pheromones have been widely studied with respect to chemistry in the orders such as Diptera, Lepidoptera, Hymenoptera, Coleoptera, Orthoptera and Hemiptera. Some important examples are given below with description and structure:

Gyplure

Gyplure is female sex pheromone of Gypsy moth *Porthetria disper*. Chemically it is 10-acetoxy-cis-7-hexa-decnol. It has a chemical structure as follows:

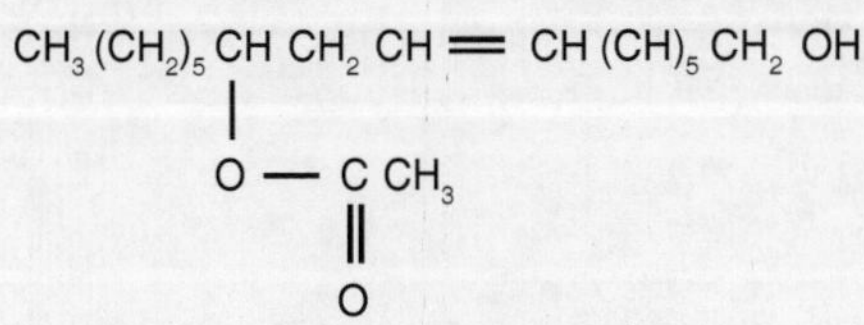

Gyplure Structure

The chemical structure of gyplure can also be expressed as:

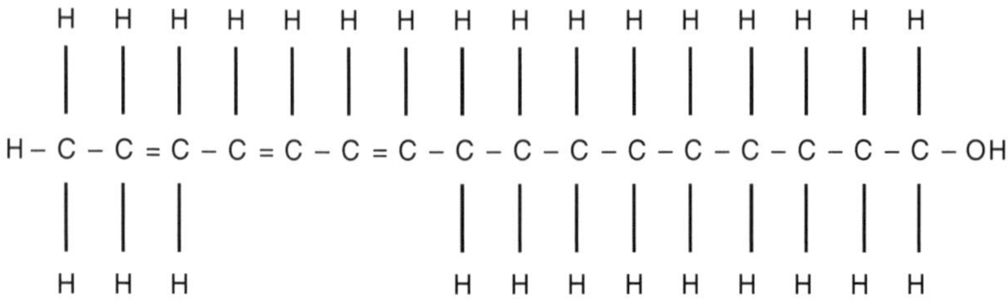

Gyplure Structure

Gossyplure

Gossyplure is pheromone produced by the female moths of the pink boll worm, *Pectinophora gossypiella*. Chemically it is 10-propyl-trans-5, 9-tridecadienyl acetate. It has the chemical structure as follows:

$$[CH_3\,(CH_2)_2]\,O = CH\,(CH_2)_2\,CH = CH\,(CH_2)_3\,CH_2O\,\underset{\displaystyle \|\atop O}{C}\,CH_3$$

Gossyplure Structure

"Gossyplure" is also referred as "Gyptol".

Looplure

Looplure is pheromone produced by the female moth of cabbage looper, *Trichoplusia ni*. It is chemically cis-7-dodecenyl acetate. Its structure is given below:

$$CH_3\,(CH_2)_3\,CH = CH\,(CH_2)_6\,OC\,(O)\,CH_3$$

Looplure

Pheromone for Indian water bug, *Lethocerus indicus*

Chemically it is "-Trans-2-hexen-1-01.acetate".

Synthesis

$$CH_3(CH_2)_2\ CHO + HO_2CCH_2CO_2C_2\ H_5 \xrightarrow{C_5H_5N}$$

$$CH_3(CH_2)_2\ CH = CHCO_2\ C_2H_5 \xrightarrow[2.\ CH_3\ Co\ Cl]{1.\ Li\ Al\ H_4}$$

Queen substance

Queen substance is released by the queen of honey bees. Mandibular gland is the source of this substance. The substance acts as attractant for drone caste of the honey bees. However, the substance affect the ovarian maturation in worker caste in the colony. Chemically it is g-oxo-2-decenoic acid. Its structure is given below:

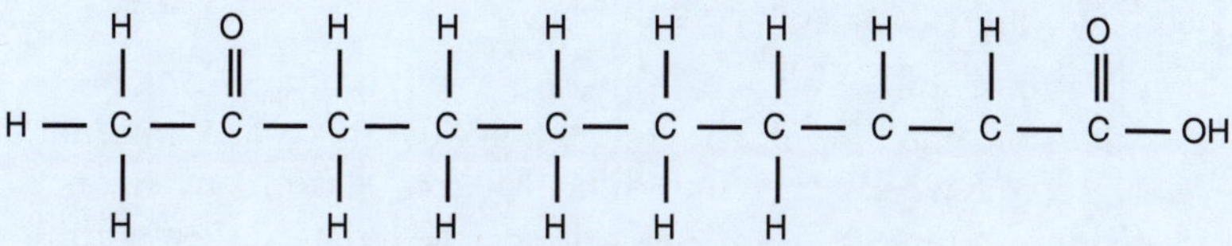

Queen Substance of Honey Bee

Almond moth (*Carda cautella*)

Cis-g, trans-12-tetradecadien-1-01 acetate

Spruce bud worm moth (*Choristoneura fumiferana*) Trans-11-Tetradecenal

The insects which depend upon pheromonal communication in their mating usually have a highly developed olfactory sense for them because, the pheromones are released in a very minute quantity. The olfactory receptors of insects are usually located on the paired antennae on the head (Jacobson, 1974). The sex pheromone in insect species is secreted under certain (physiological and ecological) favorable conditions by the secreting sex. Similarly, the sex that perceives the pheromone usually has the tendency to respond under similar conditions. This synchronization provides maximum opportunity for two sexes to come together for mating. The physiological variables include age of the insect, internal clock that indicate the time of the day, and other factors. The age of the perception of the sex pheromone in the males generally coincides with its physiologically mature age of mating. The variation in the rate of response with age of perceiving sex has been reported for many insect species (Singh & Sidhu, 1975). The perception of sex pheromone occurs at specific hours of the clock in a number of insects. The insect is able to sense the time of the clock through its daily changes in environmental cues.

The organ of perception of the female sex pheromone of *Spilosoma obliqua* (Walk) (Fig.23) is located in the antennae of the male. The males (Fig.22) antennae are broad and finely divided into lateral branches which are covered with a number of fine hair like structure, whereas the females antennae are simple and without lateral branching. The attraction of males towards the female sex pheromone is maximum at 1.00 am and then a little lesser at 4.00 am. No response is shown by the males and at day time, 24 hour aged males get attracted maximum.

Trapping pink boll worm through synthetic pheromone is possible (Siddiqui, 1988). The larval stages of this pest are internal feeders of cotton bolls. Therefore, difficult to control with conventional pesticides. In this species (*P. gossypiella*), the females use a highly potential attractant to lure males for mating, which is geometrical isomers (Z, Z and Z, E of 7, 11-hexadecadien-1-01-acetate) referred as gossyplure. The gossyplure is registered by United States Environment Protection Agency in 1978 for commercial use. Field application lure consists1 mg of 1:1 mixutre of Z, Z- and Z, E-isomers of technical gossyplure on an IF Red rubber stopper. This is found to be most attractive and economic compared to other three commercially available lures. Delta traps are most useful against *P. gossypiella* moths as compared to pherocon I.C. and Conrel kitterman. The most suitable height for fixing traps is 15 cm above the cotton plant level. Maximum catches of moths are possible in July-August. Most suitable catching time for moths is between 12.00 pm to 4.00 am.

The codling moth, *Cydia pomonella* (L.) is a serious pest of apple in J.K. The pest is controlled by sex-pheromone traps successfully. Male removal sex-pheromone traps baited with codling moth lure E, E-10, 12-dodecadien-1-01 are found widely successful in control of codling moth. The baited traps should be installed from late May to 3rd week of July and again from 1st week of August to end of August with fresh lures. One trap is sufficient for five trees.

In caster semilooper, *Achea janata* Linn. (Fig.24) females exhibit call and mating takes place between 22.00 and 00.00 hours during the day period. The natural pheromones blend of the female is identified as heneicosane; (Z_6, Z_9)-6, 9 heneicosadiene; (Z_3, Z_6, Z_9)-3, 6, 9 heneicosairiene; and (Z9, Z12)-9, 12 contadecadienal in 1.0 : 1.0 : 60-70 : 10 ratio respectively.

Helicoverpa (Heliothis) armigera is serious pest of several agricultural crops. It attacks more than 180 plants in agro-ecosystem and forest ecosystem. The discovery and synthesis of sex pheromones of *H. armigera* has paved the way to use them for monitoring and trapping the male moths. A synthetic sex-pheromone formulation, a mixture of (Z)-11-hexadecenal and (Z)-9-hexadecenal (93:3) at the concentration of 1 mg/septum is effective in moth catches.

As compared to females, very less sex pheromones are identified and characterized in males. The male pheromone components of some insect pests are tabulated in Table 4.

Table 4: Pheromonal Components in Male Insects

Sr.No.	*Insect*	*Pheromone*	*Chemical Name*
1.	Queen butterfly	Alkaloidal ketone	2,3-didydro-7methyl-1H pyrrolizin-1-one
2.	The cabbage looper *Trichopulsia ni*	—	2-phenylethanol
3.	The boll weevil *Anthonomus grandis*	Two terpene alcolhos and two terpene aldehydes	—
4.	A dipteran fly *Ceratitis capitata*	—	Methyl-6-nonenoate and methyl-6-nonenol

In a pheromone large number of carbon atoms (10-20) are noted. The pheromone substance has a high molecular weight ranging from 180-300 daltons. These substances have a narrow specificity but high potency hence, are with long carbon chain and high molecular weight.

The pheromone has multi-component system. However, first few pheromones of the pests like gypsy moth, cabbage looper, etc. are chemically characterized and found to be contain only one component. But, most recent reports indicate multi-component nature of sex-pheromones of the insect pests. Sometimes, in two components pheromones, neither compounds elicits behaviorual activity when presented individually but the mixture of two components shows optimum response.

The pheromonal activity is related to number of components present, proportion of them, age and sex of the species. In Lepidoptera there is less specificity in sex-pheromonal components. The pheromone of one species may attract many another species. However, insects have different timings for their mating and this behaviour is manageable in pheromonal pest management strategies.

The review of literature indicates that the sex pheromones have less species specificity. The chemical compounds acting as pheromones are fewer than the insect number available. Hence, it may appear that species may have limited choice of pheromones and the sympatric species may have chemically similar pheromones. However, the pheromones have different features with respect to miner structure as in the number, position and configuration of double bonds and carbon atoms. Release rate of pheromones has also the species specificity. In cabbage looper moths (T.ni) and Alfalfa looper (*Autographa californica*) Z-7 decenyl acetates acts as female sex pheromone. However, *T.ni* responds only to 10-100 times greater rate of release than that of *A. californica*. Mono and multi components are one of the important aspects of the species specificity of pheromones, by that also the insects can have their choice of pheromones to respond. Component ratio of pheromone also plays an important role in activating the moths. In *H. armigera*, a synthetic formulation of a mixture of (Z)-11-hexadecenal and (Z)-9-hexadecenal is effective with 93:3 porportion. *Adoxophyses orana* and *Clepsis spectrana* respond to (Z)-9 and (Z)-11-tetradecenyl acetates. But, *A. orana* responds to the ratio of 90:10 and *C. spectrana* responds to the ratio of 25:75. Additional minor components has also great relevance in receptivity of opposite sexes and maintaining specificity.

The review of literature suggests that a minimum of 1 molecule of Bombykol (pheromone of *Bombyx mori*) can generate a response in the antennal receptors of the male *B. mori* but, minimum 200 molecules will elicit a pheromonal behavioural response in the insect. The above range is threshold of pheromone response. The insect knows their own TPR and respond accordingly. Release of 0.01 μg of pheromone component by Gypsy moth is sufficient for stimulating 1 billion males of its species. This indicates the easiness of female to call males to her own door.

Effectiveness of pheromones with respect to distance have been studied by several workers. A saturnid male moth can respond to female from 11 km. In general, male moths get attracted to a pheromone baited trap from a distance of 2 miles or more. It is noted that the effective distance of a sex pheromone is dependent on the amount of pheromone released by the female, the threshold concentration of male stimulation and the climatic conditions.

In response with female sex pheromone, the male moth activates and start flying upwind and thus, get flying towards the female. The pheromonal concentration also helps in finding the female as there is slight but gradual increase in the concentration approaching females. Accidentally, if the moth is driven out of the active zone, then it will wander out of the zone and miss the chance or may come in active zone once again accidentally and will pick up the zone and reach to the target.

Pheromone Application Methods

Pheromone Application Methods

Old methods	*New methods*
1. Release from dishes	1. Micro encapsulation
2. Release from wicks	2. Hollow fibre
	3. Pheromone baited traps

Micro-encapsulation

Micro-encapsulation is the method of application of pheromones through plastic capsules. The capsule is very small it may casually 50 μm. It is dispersed with conventional sprayers for uniform coverage.

Hollow fibre method

This is the method of pheromone application under which the

pheromone is kept in hollow plastic fibre or capillaries. The capillaries are cut into small pieces and scattered with the help of aircraft. The capillaries may also be in hoops (coils)) and tied to upper part of the stems.

The pheromone baited traps

The pheromone baited traps are specially designed structures. They are of various shapes and size and well suited for hanging to the trees and or specially elevated polls or objects.

Sectar XC-26, Sectar-1, Pherocon R, etc. are famous pheromone traps utilized in pest management programme. Recently developed several other traps are also now commercially available in the market.

Pheromone dispensers

The pheromone dispensers are specially designed structures that can release pheromone component with appropriate and calculated rates. Hercon controlled release pheromone dispenser is widely used in pest control programmes. It is made up of plastic laminates and has three layers. Middle layer acts as reservoir for pheromone and the outer layers regulate the rate of release of pheromone component and may act as barriers for release of pheromone.

Types of Hercon Pheromone Dispensers

Following types of hercon pheromone dispensers are utilized in pest management programmes.

Square shaped

It is 0.5 to 2.5 cm in sides and can be placed manually. Few hundred squares can mark the odour of a "calling" female in one acre area which will lead to disruption in mating of the pest species.

Ribbon shaped

It is ribbon shaped and can be placed manually.

Flake shaped

Flakes are useful against forest and orchard pests and they are applied by aircraft. Experimentally it is proved that about 10,000 to 20,000 sex pheromone flakes per acre can have mating disruption effect on the pest species. They can be scattered by an aircraft to target areas.

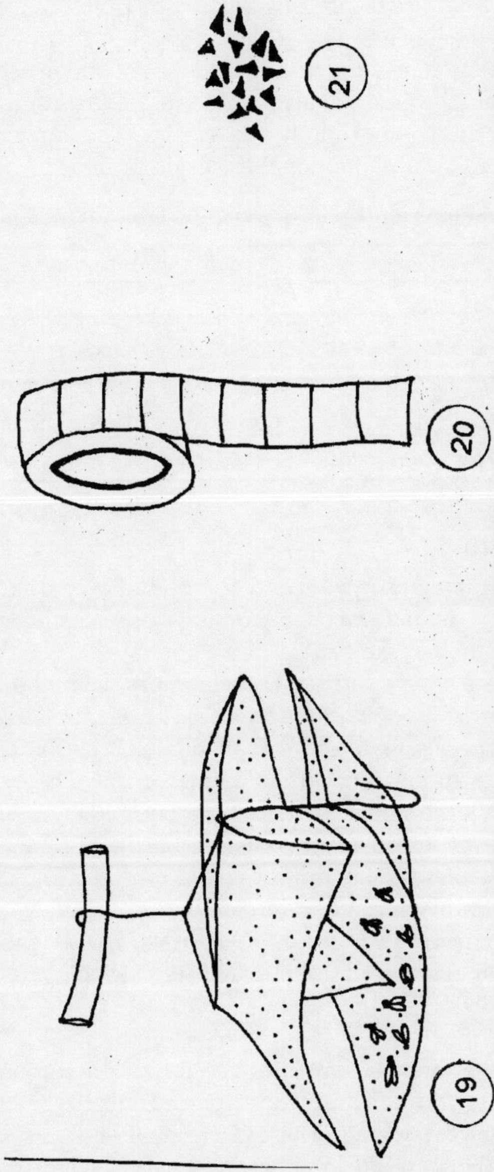

Fig. 19: Pherocon Types Pheromone Baited Trap.

Fig. 20: Ribbon Shaped.

Fig. 21: Flake Shaped.

Aggregation Pheromones

These pheromones are responsible for congregation of insect species. Congregation or aggregation of a species is aimed to protection, reproduction, feeding or combinations of above needs. Aggregation pheromones are widely studied in the insect orders such as coleoptera, dictyoptera and hemiptera, etc. The important species of insects and their pheromones are listed in Table 5.

Table 5: Insect Pests and their Aggregation Pheromones

Sr.No.	*Insect Species*	*Order*	*Pheromone*
1.	Cockroach *Periplanata americana*	Dictyoptera	Periplanone
2.	Bark beetle *Dendroctonus frontalis*	Coleoptera	Frontalin
3.	Phloem beetle *Ips confusus*	Coleoptera	Ipsenol
4.	Flour beetle *Tribolium* sp.	Coleoptera	Dimethyldecanol
5.	Pentatomid bug *Eurydema rugosa*	Hemiptera	

Alarm Pheromones

Alarm pheromones are produced for giving warning, or suggestions against fear, natural enemy or any purpose related to basic needs (food, shelter, mate). Alarm pheromones are widely studied in the insect orders such as Hymenoptera, Isoptera and Homoptera, etc. Alarm pheromones are produced by different organs in different insect species. Cornicals or bars on aphid abdomen is source of alarm pheromone in aphids. In red cotton bug, *Dysdercus* sp. abdominal terga is the source of this pheromone and in termites cephalic glands are responsible for producing alarm pheromones. While, in ants anal, mandibular, dufour's, and poison glands are visualized as a source of alarm pheromones. In case of honey bees sting apparatus and mandibular glands produce alarm pheromone. The insects and their alarm pheromones are listed in Table 6.

Trail Pheromones

Trail pheromones are also called as aerial-trail pheromones. Because, they function through the medium of air. Trail pheromones are used for finding food or mate. These pheromones have special importance in social insects like termites and ants as they are used to maintain cohesion and social integration in colony. Organs producing trail marking pheromones vary with the species. The insect species and their trail pheromonal components are presented in Table 7.

Table 6: Alarm Pheromones in Insects

Sr.No.	*Insect Species*	*Order*	*Pheromone*
1.	Aphids	Hemiptera (Homoptera)	Terpenes
2.	Other bugs	Hemiptera	Aldehydes
3.	Ants	Hymenoptera	Ethyl and methyl ketones, formic acid
4.	Termites	Isoptera	Monoterpene hydrocarbons
5.	Honey bees	Hymenoptera	
	Sting gland	Hymenoptera	Alkyl and alkenyl acetates
	Mandibular gland	Hymenoptera	2-heptanone

Table 7: Insect and Trail Pheromones

Sr.No.	*Insect Species*	*Order*	*Source*	*Pheromone*
1.	Termintes. *Zootermopsis* spp	Isoptera	Sternal gland	3,6,8-dodecatrienol
	Reticulitermes spp.	Isoptera	Sternal gland	Hexanoic, hepatanoic, decanoic, honanic, acids, etc.
2.	Ants	Hymenoptera	Tibial, Dufour's, Pavan's, Poison and Rectal gland	All combined with other components Many unknown
3.	Caterpillars	Lepidoptera	Labial gland	Many unknown

Advantages and Disadvantages

Advantages

1. Minute quantity of pheromones can kill large number of species (by attracting them).
2. They are eco-friendly, non-pollutants.
3. No adverse effects on non-target species.
4. Labour saving method.

5. Easy for monitoring populations.
6. May be helpful in quarantine.
7. Suited in IPM.

Disadvantages

1. Attract only one sex, other sex causes damage to the products.
2. Utilization needs expertise and skill.
3. Provides no quick results.
4. Not known for all species of pests.

Fig. 22: ***Spilosoma obliqua*** **(Male Moth).**
Fig. 23: ***S. obliqua*** **(Female Moth).**
Fig. 24: ***Achea janata*** **Moth.**

Chapter 5

HORMONAL PEST CONTROL

The physiological processes related to reproduction, growth, moulting and metamorphosis in insects are regulated by the hormones. Thus, control of hormones is control of insects. The quality, quantity of hormone and time and stage of an insect play an important role in successful development of an insect. Interference in above things will lead to no development/intermediate stage of the insect/improper development which ultimately results in death of the insect.

Hormones Involved in Pest Control Strategies

Brain hormone, juvenile hormone and ecdysone hormone produced by endocrine organs have important role in pest management. Other hormones produced by an insect have no role in pest management hence, not discussed here.

Brain hormone

Brain hormone is also called as neuro-hormone which is synthesized by neuro-secretary cells (NSC) of the brain. Brain hormone is also visualized by activation hormone (AH). Brain hormone is responsible for activation of carpora allata(CA) and prothoracic gland (PTU) to release their own hormones. Allatotropin activates CA while, prothoracicotropin activates PTU.

Juvenile hormone (JH)

It is produced by carpora allata by activation of allatotropin hormone of brain. It is non-steroid lipid of terpenes group. General structure of JH is given below:

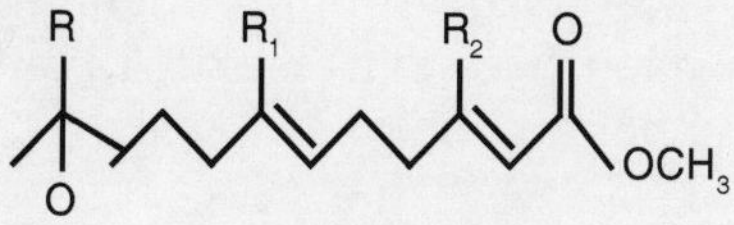

JH

There are four types of JHS structurally different from each other namely, JH-0, JH-I, JH-II and JH-III. The structures of which are given below:

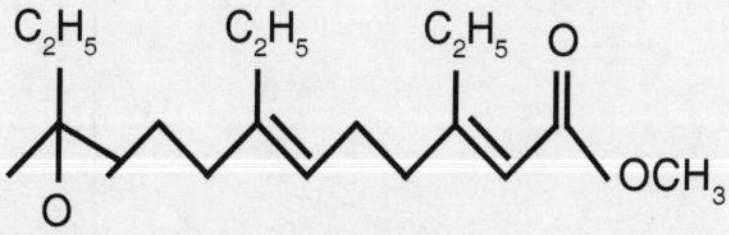

JH-0

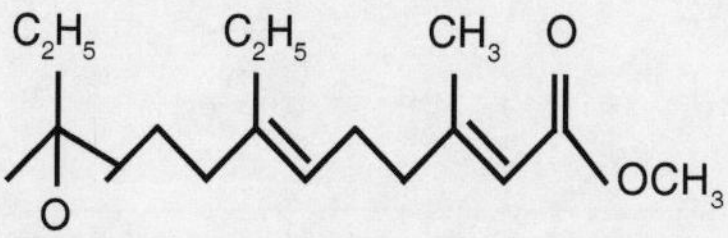

JH-I

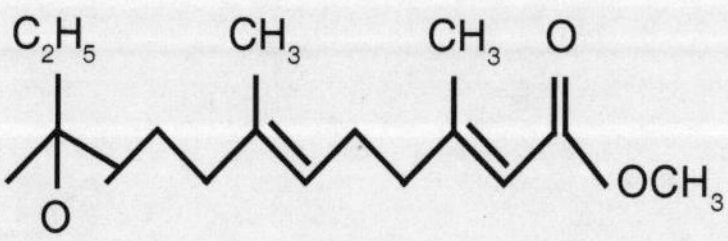

JH-II

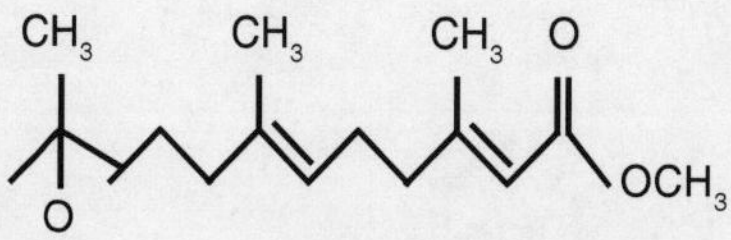

JH-III

JH-I and JH-II have been identified in all Lepidopterous species studied and JH-III is reported from all the orders studied so far. However, JH-0 is reported only from the moths of *Manduca sexta*. The hormones JH-I and JH-II are morphogenic while, JH-III is gonadotropic in action. In certain cases JH-III can act in both ways in absence of other hormones.

Ecdysone hormone

It is produced by prothoracic gland. Chemically it is steroid in nature. Ecdysone is also referred as moulting hormone (MH). It contains two hormones *i.e.* α-ecdysone and β-ecdysone. α-ecdysone is prohormone which can be converted into β-ecdyone in the peripheral tissues of the body. β-ecdyone is the real moulting hormone since it brings about moulting in insects. It is also familiar by other names, ecdysterone, crustecdysone and 20-hydroxyecdysone.

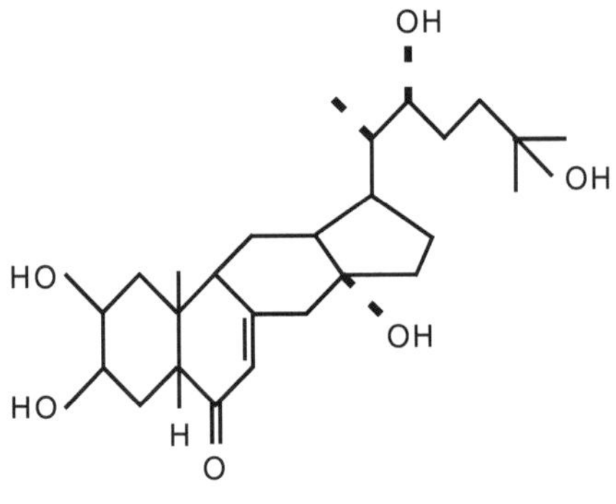

α – Ecdysone Structure

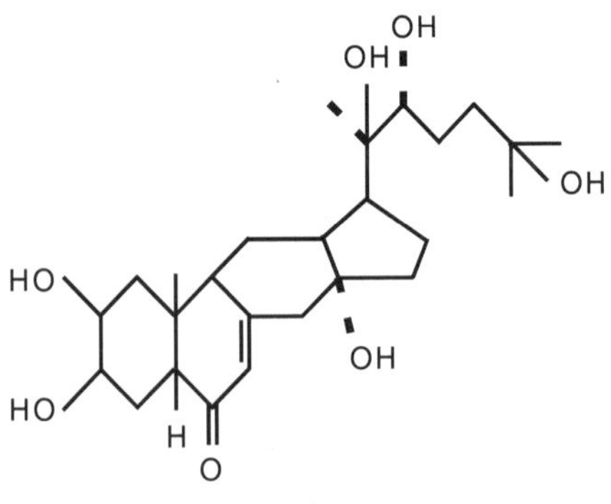

β – Ecdysone Structure

The brain is stimulated by the environmental factors such as day light, temperature, humidity, etc to release Activation Hormone (AH). AH then stimulates prothoracic gland to produce MH and carpora allata to produce JH. The MH initiates moulting process and takes part in periodic casting of cuticle. JH hormone governs the rate of maturation including the development and maturation of gonads. In higher titer of JH, larva moults into larva but in low titer it moults in pupa. However, JH is not involved in the adult formation from pupa. In hemimetabolus insects also, the last instar larva moult directly into adults in absence of JH. MH has no role in adult stage of an insect. Moulting is stopped in adult stage, therefore, PTU disappears in early part of the adult. However, CA is activated once again for release of the hormone gonadotropin which is needed for the maturation of ovary.

The main function of juvenile hormone in insect development is to determine the result of each moult by suppressing the expression of adult characteristics till the appropriate time. If JH is provided exogenously (outside) to last instar, it will fail to develop into pupa and likely pupa into adult. In fact, larva will land into intermediate supernumerary stage (extra moult) and pupa will form pupal - adult intermediate called an "Adultoid". These intermediate stages, however, can not stay for longer period but die soon.

Pest Management Strategies

According to Williams (1956) JH hormone can be used to kill the insect. Thus, insect hormones are recognized as "third generation pesticides". JH mimics are found in certain plants. From balsam fir tree (*Abies balsamea*) "Juvabione" is extracted. From beeding plant (*Ageratum houstonianum*) prococenes I and II have been collected. These products destroy carpora cardiaca and thus act as inhibors of JH synthesis. Hormones regulate insect development and metamorphosis.

Anti-hormones disrupt hormonal processes in insects and cause death of the individuals directly or indirectly. Thus, anti-hormones play a very important role in control of pest populations. Therefore, anti-hormones are called as "fourth generation pesticides".

Insect growth regulators (IGRS) are synthetic analogues of hormones which regulate insect development and metamorphosis. Non-hormonal compounds are also investigated recently which interfere with the development of insects and cause mortalities in them.

Insect hormones are chemically characterised and their mimics (analogues) have been synthesized. For increasing effectiveness of analogues their chemical structure is slightly changed. Collection of natural insect hormones and their application in the pest management is very difficult since insect hormones are produced in very less quantities and they are thermolabile. Hence, synthetic analogues are of great importance for the use in pest control programme. Ecdysoids and juvenoids are the synthetic analogues of ecdysone and JH respectively and can be used in insect pest management.

Pest Management through Ecdysoids

The ecdysoids are responsible for killing the insect pests. They penetrate through cuticle in most of the insects. Hence, they are applied to pest species topically in various solvents. Some ecdysoids are applied through diet as stomach poison while, some are synergised by mixing with piperonyl butoxide and used as conventional insecticide. The ecdysoids are responsible for formation of defective cuticle in the pest insect and thus kills them. The ecdysoids also kill the insects by juvenilizing effects. By exposing insects to large amount of an artificial hormones the developmental processes can be accelerated with some normal events and some abnormal events like lack of the uppermost wax layer in cuticle or lack of scales on integuments. This will result in death of insects due to lack of protection and desiccation. Likely, low concentration of hormone in normal development will help DNA for slow and gradual re-programming in genes to produce the cuticle of the next stage but under high concentration of ecdysoid the DNA will not get sufficient time to do so and it will continue to secrete the cuticle of the same stage and keep the insect in a juvenile condition only, which can not survive longer. The excess of ecdysone hormone is referred as "hyper ecdysonism" or "hyper hormonism".

Moulting Hormone derivatives (MHd)

MHd can be divided into four groups namely,

1. Zoo ecdysoids,
2. Phyto ecdysoids,

3. Synthetic ecdysoids, and
4. Anti ecdysoids.

Zoo ecdysoids

For insects, ecdysterone is an important zooecdysoids. It is more active than ecdysone. Chemically, it is 20-hydroxy ecdysone. In some insects like locusts, it is inactive but, ecdysone occurs in large amounts in grasshopper faeces. In addition to this, several other steroid hormones have been reported in insect pests like Tobacco horn worm (*Manduca sexta*), etc.

Phytoecdysoids

These are plant steroids with insect moulting hormone activity. From the leaves of the fern (*Plodocarpus nakaii*) four substances closely related to ecdysone and ecdysterone were isolated (Nakanishi, *et al.* 1966). Later, Takemoto, *et al.* (1967) added two steroids viz. Inokosterone and Ino-inokosterone from the roots of fern (*Achyrantes fauriei*) and one more substance, polypodine from common european fern. These substances are more effective than ecdysterone. Since then number of members have been added to this group. More than 100 phytoecdysoids have been reported and isolated. Ponasterone A and Ino kosterone are also found in insects like blow fly (*Calliphora stygia*). Several new steroids have been isolated from a variety of plants which showed MH activity (Russel, 1972).

Synthetic ecdysoids

A large number of steroids with MH activity are investigated and tested against the pest like *Calliphora*. 22-isoecdysone is investigated by Mori *et al.* (1968) and 20-hydroxysteroids by Middleton *et al.* (1972). Slama *et al.* (1973) reviewed the synthetic ecdysoids.

MH Inhibitors

More than 20 plant extracts showed anti-ecdysone actions in *Chillo* larvae. CL-A (plant extract) inhibites the MH activity of ecdysterone strongly than ponasterone A. Four types of azasteroids are potent inhibitors of growth and development in insects such as flies, beetles and cockroaches.

Goods of MHd

1. They are active in extremely low concentrations.

2. Their action is not limited to specific stages and periods of insect development.
3. Inactive in other animals.
4. More stable than juvenoids in open environment.

Bads of MHd

1. Poor capacity of penetration through cuticle.
2. Less developed than juvenoids.

Pest Management by Juvenoids

Juvenoids are the most effective pest control agents among the hormonal insecticides. They cause mortalities in insect pests in following way -

1. As larvicide,
2. As ovicide,
3. As anti-metamorphic,
4. By inhibiting embryogenesis, and
5. By disrupting the diapause.

Recently, several chemical compounds have been discovered which shows larvicidal effect in insect pests. The compound, altosid has larvidical effect on mosquitoes. This compound has been registered under the trade name "methoprene" for use in pest management in USA. It has been also reported from other several countries as an effective larvicide for control of mosquitoes. For causing mortalities in mosquitoes the compound methoprene is treated to ponds, pools, etc. This juvenoid produces derangements in the larval development, thereby killing the larvae. Methoprene effectively controls the larval forms of horn fly *Haematobia irritans*, stored tobacco pests belongs to coleoptera and lepidoptera, certain greenhouse homopterans, dipterous leaf miners (Agromyzidae) damaging vegetables and flower crops, etc.

JH is needed for maturation of ovary but, in small quantity. However, eggs of the insect pests will be made inviable or sterile by inducing a large exogenous dose of juvenoid to adults. By this way egg production can be stopped. Eggs will be made sterile by mating of females with juvenoid treated males.

Juvenoids shows anti-metamorphic effects on insect pests. It prevents metamorphosis and forces the larva to continue as a larva only. Likely, when the juvenoids are provided exogenously, the larva will go under extra moult or moult into defective intermediate form which may suffer from a failure to successful moult, feed or mate and by that affected to death. Thus, presence of the hormone at wrong time with wrong quantity will lead to the death of the pest insects.

Juvenoids can interfere with the embryogenesis of an insect by affecting development of embryo and also further may act as anti-metamorphic component in post embryonic development. The embryo of an insect will not complete its development when the egg is treated by hormone and if it develops and hatch, the hormone may persist to reveal its effect at a later stage, specially, the last larval instar may fail to metamorphose into an adult.

Juvenoids can be used to disrupt the diapause of insect pests. Diapause is an essential adjustment of the insects against environment. Diapause may be disrupted in two ways. Juvenoids can terminate pupal diapause by activating the inactive PTG of a diapausing pupa or preventing pupa from entering into diapause. In both cases the insect will die due to environmental or climatic stress.

JH Analogues

More than 1000 JHa substances have been reported. Eight types of JHa's are mainly recognized in pest management of insect species. These are listed below:

1. Substances similar to the active component of the abdominal extract from adult.
2. Acrylic terpenoids.
3. Aromatic ethers, thio ethers and amines, etc.
4. Monocyclic sesquiterpenoids.
5. Aromatic derivatives of juvabione.
6. Acyclic compounds of the dodecyl ether type.
7. Certain fatty acids.
8. Recent compounds with peptide structure.

The insect pests tested against JHa substances are tabulated in Table 8.

Table 8: Insects Tested Against JHa Substances

Sr.No.	*Species Pest*	*Substance*
1.	*Pyrrhocoris* sp.	L-isoleucyl-L-alanyl-aminobenzoic acid ethyl ester
2.	*Papilio demoleus*	Oleic, linoleic, erucic acid in aqueous emulsions
3.	*Musca domestica*	Esters of saturated and unsaturated fatty acids
4.	*Ephestia cautella*	—
5.	*Galleria mellonella*	Abdominal lipids
6.	*Dermestes maculatus* (adult)	Abdominal lipids
7.	*Tenebrio molitor* (larva)	Abdominal lipids
8.	*Aedes aegypti*	Crude synthetic JH preparation
9.	*Choristoneura fumiferana*	JHa topical application
10.	*Laspeyresia pomonella*	JHa topical application (+ve effect)
11.	*Danaus plexippus*	JHa topical application
12.	*Malacosoma pluviale*	JHa topical application
13.	*Pieris brassicae*	JHa topical application
14.	*Polistes metricus*	JHa topical application
15.	*Leucophaea maderae*	JHa topical application
16.	*Pyrrhocoris* bug	JHa topical application (-ve effect India)
17.	*Erias fabia* (Fig. 26)	Crude juvebione
18.	*Spodoptera litura*	JHa (-ve)
19.	*Pectinophora gossypiella*	JHa (-ve)

Pest Management by Antihormones

Antihormones are those compounds which interfere with normal activities of hormones and cause mortalities in insects directly or indirectly. Antihormones are referred as "Precocenes". They may function as anti-JH. They may destroy carpora allata and thus prevent JH synthesis. Precocenes treated insects produce tiny precocious adults which can not mate or can not lay eggs and also can not survive longer but, die soon. Precocenes affect insect pests by inhibiting ovarian maturation, by inhibiting JH production and inhibiting certain steps in the synthesis of ecdysone.

Antihormones are found effective mostly against heteropetrous pests and certain coleopeterous pests only. Recent new compounds like EMD, FMeV, PB etc. have been investigated as antihormones for insect pests in different groups. EMD (ethyl-3-methyl-2-dodecenoate) competes with JH receptors and inhibits the action of endogenous JH; FMeV (Fluoromevalonolactone) inhibits early steps in natural JH synthesis while, PB (Pyperonyl butoxide) inhibits terminal steps in natural JH biosynthesis.

The new compounds such as tripamol and azasteroids have also great relevance in pest management since they inhibit certain steps in the synthesis of ecdysone and affect normal development of the insect.

Pest Management by Chitin Synthesis Inhibitors

Chitin is essential component of body wall of an insect. Chitin synthesis is essential for periodic shedding of cuticle. Chitin synthesis is to be prevented for preventing the casting of exuvium and causing the death of an insect. Chitin synthesis inhibitors have been investigated in 1970 in the form of Benzoylphenyl ureas. They interact with enzyme Chitin synthetase and block its activity in vivo. Many compounds from natural products such as plumbagin and antibiotics such as polyoxins and nikkomycins and insecticides like benzimidazole and fungicides like captan have shown to block chitin synthesis in insects. However, benzoylphenyl ureas, the newest compounds play an active and important role in pest control by inhibiting chitin synthesis. Chitin synthesis act mostly by interfering with deposition of chitin. The treated larvae have difficulties in moulting. The larvae are unable to cast the exuviel and finally die. These compounds also act as ovicides and larvicides in insects.

Chitin Biosynthesis Pathway

Chitin is a haemopolymer of N-acetyl-glucosamine which occurs predominantly in the procuticle of insects, synthesized from glucose. Uridine diphopho N-acetyl glucosamide forms the precussor of chitin microfibrils of varying length and are covalently linked to protein. The cuticular chitin is degraded along with protein at the time of moulting by chitenase enzyme. The chitin biosynthetic pathway of gram pod borer is shown below:

Glucose

↓

Glucose-6-Phosphate

↓

Fructose-6-Phosphate

↓

Glucosamine-6-Phosphate

↓

N-Acetyl Glucosamine-1-Phosphate

↓

Unidine-diphospho-N-Acetyl Glucosamine
polymerized

↓

Chitin Synthetase

↓

Chitin

The newly formed cuticle without synthesis of chitin, will not produce chitin in cuticle. Thus, during moulting the insect will be unable to withstand muscular contraction and will prevent the casting of exuvium and finally die. As per the recent views of scientists it is believed that the moult inhibitors may prevent the transport of UDP-AGA across to biomembrane resulting in prevention of deposition of chitin either in the cuticle or peritrophic membrane of midgut.

There are several moult inhibitors released by commercial companies. PH-60-38 is released by Philips-Dupher, BAY SIR 6874 by Chemagro Limited and I-1215 by Eli Lilly & Co. CME-134, XRD-473, IKI-7899 and DP-145 are also important moult inhibitors available in the market which are manufactured by Celamerk CmbH, Dow Chemical Co. Ltd., Ishishara Sangyo and Kevisha Ltd. respectively.

Following three groups of compounds are visualized as moult inhibitors/chitin synthesis inhibitors.

Diflubenzuron

It is also called as "Dimilin". It contains two halogens, chlorine and fluoride. The benzene rings are connected by urea bridge.

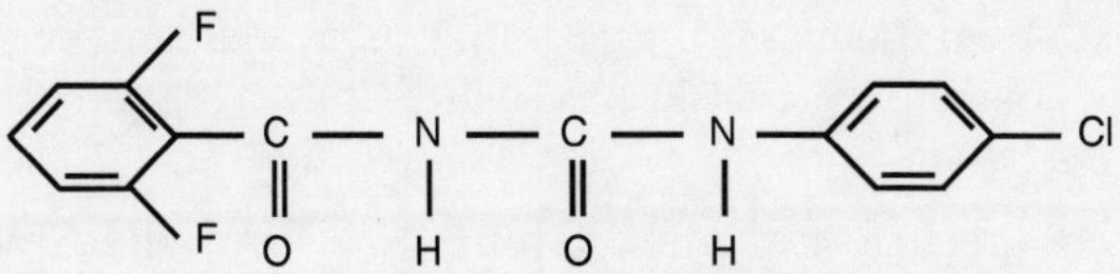

Diflubenzuron

Diflubenzuron compounds available in market are mentioned below:

PH-60-40, PH-60-41, PH-60-43, etc.

Penfluron

It is structurally similar to diflubenzuron except Cl is replaced by CF_3.

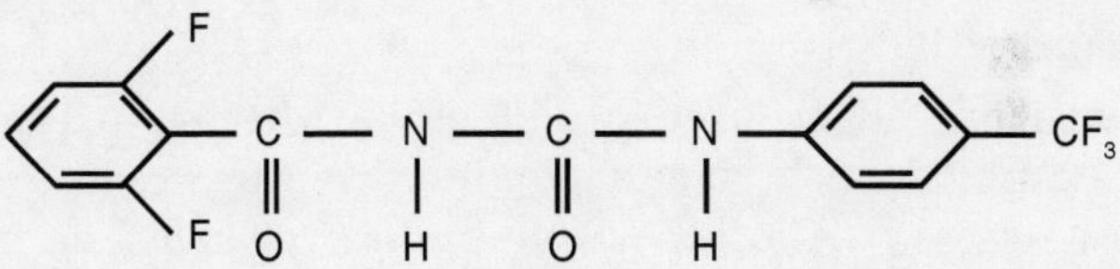

Penfluron

Other examples are: PH 60-44, PH 60-45, etc.

Triflumuron

Following are the important compounds available under triflumuron group which disrupt moultings in insects.

Example: BAY SIR 8514, BAY SIR 6874, etc.

The above compounds interact with insects by disrupting moulting, failure to feed, delaying symptoms of characters of the stage, ovicidal effects, larvicidal effect, contact effects, etc.

The above compounds are marketed in more than 15 countries in the world for use in pest management and known by inhibiting moulting of the insect pests of agriculture, forest, horticulture, sericulture, fibre crops, household pests, stored grain pests, etc. The most significant characteristics of these compounds is that they are safer for natural enemies of the pests (like parasitoids and predators) and even pollinators (insects) hence, worthwhile in IPM programmes of crop pests.

According to Retnakaran & Smith (1975) Benzoylphenyl ureas have ovicidal or embryocidal, larvicidal and other minor effects on insects. Most insects show malformation of the cuticle depending upon dosage, species and the compounds (B.U.). In some pests contact effects are noted. In *Helicoverpa* (=*Heliothis*) *armigera* DFB had significantly reduced the egg hatch. For *H. armigera,* among the several methods of treatment *viz.,* ingestion, dipping, topical spray and contact method, ingestion is superior in causing mortality in the larvae. DFB is also effective against spruce bud worms, *Spodoptera litura* (Fig. 25) and many more insect pests. According to Balasubramanium (1990) DFB fits well in IPM programmes since it has different modes of action and ecological acceptability with insecticides and fungicides.

Miscellaneous IGRS

Several miscellaneous IGR products have been tested against specific pests under field conditions with promising results. Zoecon company released "Altocid" which is effective against horn flies and stable flies. It adversely affects the development of the pest insects. Altocid is also effective against mosquitoes. The growth regulator "Enstar 5-E" is also available in market for control of hemipterous insect pests like aphids and white flies. Prostaglandin, a vertebrate hormone is effective against red cotton bugs *Dysdercus* spp. It reduces egg hatch and interfere with fecundity and fertility of red cotton bugs. Likely, caffeine and aminophylline inhibit adult development in *Cercropria* moths.

Advantages and Disadvantages

Advantages

1. Method is economical and ecofriendly.
2. Highly species specific to apply in the field.

3. Rightly fits in IPM programmes.
4. Non pollutant, highly degradable and non-persistant residues.
5. Non-toxic to man, animals and plants.
6. Effective method of pest control for short life cycle pests.

Disadvantages

1. It needs specific stage of pest and period hence, cannot be applied at all times.
2. Pests may develop resistance against these compounds.
3. It is new method of pest control, needs popularization among the farmers.
4. Affective at last instars hence, early stages of the pest may cause damage to crops.
5. Most JH analogues are unstable, thermolabile in the environment hence efficacy is adversely affected.
6. Limits commercial viability of the method.

Fig. 25: ***Spodoptera litura*** **(Moth).**
Fig. 26: ***Erias fabia*** **(Moths).**

Chapter 6
ATTRACTANTS

Conventional pesticides cause several problems like pest resistance, secondary pest outbreak etc., hence the use of attractants is worthwhile strategy in pest management programmes. The use of attractants is behavioural manipulation of the insects and is sophisticated but simple method of pest control. Dethiers (1947) for the first time published a monograph on insect attractants. His publication stimulated several workers in the world for research. Later Beroza (1960) contributed on insect pest attractants. Likely Brown *et al.* (1961) investigated chemical attractants for housefly. Similarly, Ihoue & Ohno (1961), Jacobson (1962, 1963), Jacobson & Beroza (1963, 1964), Perez & Long (1964), etc. also reported attractants for several insect pests.

Definition

Attractants are the substances of pest management which attract the insect species towards target objectives in want of mate, food and shelter. The attractants influence the gustatory (taste) and olfactory receptors.

Classification of Attractants

ATTRACTANTS

Natural

- A. *Animal Origin*
  1. Pheromones
  2. Decomposing products
  3. Other substances
- B. *Plant Origin*
  1. Floral scents
  2. Essential oils
  3. Secondary metabolic plants parts
  4. Acids
  5. Others

Synthetic

- Sex attractants
- Other synthetic attractants

Pheromones

Pheromones are ectohormones directly released in the atmosphere. They are exocrine substances which receive specific response in individual of the same species. It may be a single chemical compound or a mixture of chemicals. In insects following types of pheromones have been identified:

1. Sex pheromones,
2. Alarm pheromones,
3. Aggregation pheromones, and
4. Trail marking pheromones.

The sex pheromones control mating behaviour and thus population of pest species. The following table shows some examples of pests and their pheromones.

Table 9: Pheromones of Insect Pests

Sr.No.	*Species*	*Pheromones*
1.	Codling moth	Trans, trans-8, 10-dodecadein-1-ol
2.	Cabbage looper	Cis-7-dodecenyl acetate
3.	European corn borer	Trans-11-tetradecenyl acetate
4.	Gypsy moth	Cis-7,8-epoxy-2-methyl octadecane
5.	Oriental fruit fly	Cis-8-dodecenyl acetate
6.	Pink boll worm	Cis-Cis-and cis-trans-7,11-hexa decadenyl acetate
7.	Southern army worm	Cis-9-tetradecenyl acetate

Details of pheromones are given in Chapter 4.

Decomposing Products of Animals and Other Substances

Scavenger beetles are attracted to dung. Many blow flies identify dead bodies within first 24 hours and use such bodies for egg laying and larval development. Fish meal is very effective attractant to sorghum shoot fly (*Atherigona socata*) on jowar ecosystem. The flies are caged in a container perforated by special exits containing fish meal powder as a IPM device developed by NARI, Phaltan, Maharashtra. More research is needed on attractants of decomposing products. Carbon dioxide, lactic acid and water are important attractants for blood sucking insect pests. Animal perspiration, blood and urine are all attractive to mosquitoes (*A. aegypti*), in varying degree, and that attractiveness is increased several times by a high concentration of moisture and carbon dioxide. Amino acids such as lysine, alanine and to a lesser extent, arginine are all attractive to the mosquitoes. It is further believed that the attractiveness of lysine may be due to its ability to act as carrier for carbon dioxide. Estrone and estradiol are responsible for attracting mosquitoes more towards women than men. Aqueous solution of 5 per cent malt extract, 0.5 per cent ethanol, 0.02 per cent skatole and 1 per cent acetal is most attractive to adult houseflies. The above solution is equally effective against *Drosophilla.* Cis-3-hexen-1-ol is a good attractant for termites, *Kalotermes flavicollis* and *Microcerotermes edentatus.* The substance is isolated from sternal glands and nest galleries of termites.

Plant Products

Several plant products stimulate olfactory receptors of insects and insects get attracted towards the source. Floral scent is good attractant for many nector feeding insects. Essential oils present in the plant are also effective attractants for several phytophagus insects. Seeds of cruciferae, onion plants, some plants of umbelliferae, flowers of *Araujia sericofera,* bark of some trees are good sources of attractant substances in plants. Plant origin attractants are listed in Table 10.

Table 10: Some Plant Natural Products as Insect Attractants

Sr.No.	*Insects*	*Plant Product Lure*
1.	*Hylemya antiqua*	Proplymercaptan from onions
2.	Carrot fly *Psila rosae*	Phenylpropanoids from umbelliferae
3.	Lepidopterous pests	Phenylacetaldehyde from flowers of Araujia
4.	Pests of cruciferae	Isothiocyanates from seeds of cruciferae
5.	Bark beetle	Terpenes from bark in combination with pheromones

Wood partially decayed by *Lenzites trabea* (fungi) is highly attractive for termites, *Reticulitermes flavipes* and other subteranean termite species. An aliphatic unsaturated ketone, oryzanone is quite attractive to the rice stem borer, *Chilo suppressalis* (Walker). The above compound is isolated from rice plant only (Munakata *et al.,* 1959). Pine beetles (*Cryphalus fulvus* and *Blastophagus piniperda*) get attracted to red pine bark as it contains benzoic acid. A spruce bark beetle, *Ips typographus* is highly responsive to a mixture of linseed oil and methyl esters in fields. Guanosine monophosphate and several amino acids like lysine, leucine, isoleucine and methionine are the active components of casein which acts as good attractant for housefly. A mexican bean beetle, *Epilachna varivestis* found responsive to the substance which is a mixture glucose, glucoside and unknown component (Nayar & Fraenkel, 1963). Coumarin acts as an important attractant for the vegetable weevil, *Listroderes costirostris obliquas*. Glycosides are responsible for the attraction of host plants to larvae of Tobacco horn worm, *Manduca sexta* and colorado potato beetles. The extracts of milkweed seed coats, elm bark and corn kernals, silks, leaves and roots are effective attractants for the adults of large milkweed bug (*Oncopellus fasciatus*), European elm bark beetle (*Scolytus multistriatus*) and corn root worms (*Diabrotica longicornis* and *D. virgifera*) respectively (Jacobson, 1966).

Synthetic Attractants

Sex Attractants

Sex attractants have been investigated for several insects pests (Table 9). 'Gyplure', d-12-acetoxy-cis-9-octadecen-1-ol have been commercially synthesized in a good yield. It is as attractive (10^{-5} μg) as the natural lure (10^{-7} μg) in field traps. Natural lure of *Porthetria dispar* is attractive to males of two species of Gypsy moth, *Porthetria dispar* and *P. monacha* but artificially synthesized Gyplure is only effective against *P. dispar*.

Other Synthetic Attractants

A series of esters of 4 (or 5) -chloro- and bromo-2-methyl-cyclohexane carboxylic acid have been tested against Mediterranean fruit fly, *Ceratitis capaitata*. The most effective attractants among the tested products are "medlure" and "Trimedlure", secondary and tert-butyl esters respectively. Trimedlure commercially is very effective product. For male melon fly, *Dacus cucurbitae* 119 compounds have been tested but the best attractant commercially available is "cue-lure". Chemically it is 4-(p-acetoxyphenyl)-2-butanone. Pine bark beetle attractants such as n-propyl benxoate, isoamyl benzoate, O- and P-toluic acids, 2-hydroxy-3-methyl benzoic acid, linolenic acid and palmitic acid, methyl linolenate, dipentene, 2-pentene and terpentine oil are highly active compounds. Many attractants are detected for *Milichiella lacteipennis* in the form of lower aliphatic amines, however, the best ones include ethyl-, n-prophy-, and isopropylamine. For the termite, *Reticulitermes flavipes*, very effective attractants are investigated which includes, cinnamyl alcohol and isosafrole, d-camphor, cinnamyl acetate, cinnamyl cinnamate, 3-methyl-α-ionone, 2-methyl-2-hepten-6-one, and N-crotyl-P-toluidine. For japanese beetle, *Popillia japonica* efficient attractants are synthesized as geraniol, eugenol and phenthyl propionate. An effective synthetic attractant is also available for coconut rhinocerus beetle, *Oryctes rhinoceros*, chemically it is ethyl-3-isobutyl 1-2, 2-dimethyl cyclopropane carboxylate. Likely, methyl eugenol and heptyl butyrate are important synthetic attractants for oriental fruit fly, *Dacus dorsalis* and yellow jacket wasp, *Vespula* sp. respectively. Commercial citronella oil, aliphatic acids (caproic acid and valeric acid), USP eugenol, some esters are very effective attractants for Japanese beelte *Popillia jaonica* investigated by USDA in 1969.

Table 11: Attractants for Insect Pests

Sr.No.	*Type*	*Attractant*	*Species*
1.	*Oviposition lure*	(a) P-methylacetophenone	Rice stem borer
		(b) Extract of corn	*Heliothis armigera*
2.	*Natural food lure*	(a) Floral scent	Many insects
		(b) Essential plant oils	Many phytophagus insect pests
		(c) Flowers of *Catasetum Cycnoches* and *Stanhopea* produce mixture of 12 terpenoids	Males of Euglossa bees
		(d) Mulberry leaves	*Bombyx mori*
		3-hexen-1-ol	Young larva
		2-hexen-1-ol	Older larva
3.	*Poison baits*	(a) Moistened bran base+ molasses+insecticides (Paris green, Sodium fluosilicate, DDT)	*Biting insects*: Crickets, caterpillars, grasshoppers, etc.
		(b) Sugar+insecticides	*Lepidoptera*: Moths and butterflies

contd...

Table 11–contd...

Sr.No.	*Type*	*Attractant*	*Species*
		(c) Trimedlure+cuelure+ methyl eugenol	*Fruit flies*: Oriental, melon and Mediterranean
		(d) Thallous sulphate+ peanut butter mixed thoroughly	Meat loving ants
		(e) Fresh vegetable pieces or germinating seeds+insecticide (planted 3 ft deep.)	Subterranean insects
		(f) Thallous sulphate+sugar+ honey+glycerine+water	Sweet loving ants
		(g) White or yellow phosphorus+a sweet syrup converted into pellets or granules	Cockroaches and rats
		(h) 2-3% metaldehyde, a crystalline polymer of acetaldehyde	Garden snails and slugs

Advantages and Disadvantages

Advantages

1. Attractants are insect group specific hence do not affect other organisms.
2. They do not kill insects hence it is ecofriendly pest control
3. Residues does not last longer in environment hence cause no pollution or less pollution.
4. They can be used for mass trapping of insects.
5. Suitable for application in IMP.

Disadvantages

1. Insects has ability to avoid treated surface hence attractants have limitations.
2. Cannot solve the problem of pest control alone.

Chapter 7
REPELLENTS

During last fourty years there has been enormous increase in the use of pesticides which has resulted in water and air pollution, health hazards, development of pest resistance, secondary pest-outbreak, pest resurgence, etc. The above facts indicates that there is need to find out an alternative for chemical pest control. At least the use of pesticides should be minimized. The use of repellents is one of the strategy for avoiding/minimizing the use of conventional pesticides. Dethiers (1947) first contributed a monograph on insect repellents and he reviewed the subject in 1956. Jacobson (1966) added great relevance on the topic through literature and research. Since then several workers attempted the insect repellents and published fragmented papers.

Definition

Repellents are the chemicals used in pest management which prevent insect damage to plants or animals by rendering them unattractive, unpalatable or offensive. Repellent substances are of great importance in pest management since they are precise or specific and ecologically sound. Repellent chemicals and mixtures protect an individual from the bites of various pests. Hence, more than one thousand of candidate compounds have been tested in the world and some of them are found effective against more than 15 species of insect pests.

Repellents are classified as follows:

REPELLENTS

- *Chemical Repellents*
  - 1. Natural
    - A. Animal Origin
    - B. Plant Origin
  - 2. Synthetic
- *Physical Repellents*

Natural Repellents

Natural repellents may be from animal origin or plant origin.

Animal Origin Repellents

Insects produce mainly aldehydes and acids like Formic, Benzoic, Tiglic, Caprylic, Methacrylic, etc. as a defence substance. These aldehydes and acids are simple in nature and of short chain length as compared to the relative complexity of the quinones produced by other arthropods. Aldehydes are acting directly on ants as repellents (Remold, 1963).

Plant Origin Repellents

Penfold & Morrison (1952) reported 40 essential oils as repellents for various insect pests like mosquitoes, sand flies, march flies etc. The most effective were those of Huon pine wood (*Dacrydium franklini*) and *Zieria smithi*. Mayer (1952) tested many essential oils as repellents for several species of mosquitoes. He found very satisfactory results from the oils of *Eucalyptus* and *Caryophyllum* against several species of mosquitoes. Perttunen (1957) extracted α-pipene from the plants, *Hylurgops palliatus* and *Hylastes ater*. He found that α-pinene derived from *H. palliatus* was more effective than *H. ater*.

Dendroctonus beetles are effectively repelled by saturated resin vapours of its host plant, pine. Likely, the adults of boll weevil, *Anthonomus grandis* are successfully repelled by the volatiles obtained from cotton seedlings. Atleast 17 species of insects can be repelled by the vapours of nepetalactone (Catnip).

In early days strong pungent odoured plant extracts were in use for repelling the insect pests. The extract was rubbed on the clothing or skin or hanged from a rafter. The oils extracted from citronella, camphor and cedar wood were important repellents for mosquitoes and other household insects. Citronella is famous even today by "Odomos" repellent. The Lemongrass (*Cymbopogon nardus*) is good source of repellent oil which contain following components: citronellol, citronellal, geraniol, boreneol, terpenes, etc. Out of which first two have strong repellent action against the mosquitoes. Pyrethrum extracted from *Crysanthemum cinerariaefolium* has also good insect repellent action. In low concentrations it acts as good repellent for mosquitoes, sand flies and veterinary pests. It has been tried in African countries against tsetse fly, *Glossinia morsitans*. One can get rid from the bites of mosquitoes like *Anopheles, quadrimaculatus* and *Aedes aegypti* and stable flies like *Stomoxys calcitrans,* etc. due to the above repellents. The plants contain chemicals like terpenes, phenols, quinones etc. These substances may be utilized in pest management as pest repellents.

Synthetic Repellents

Important synthetic repellents are given below:

Bordeaux Mixture

It has good repellent action against hemipterous and coleopterous pests. However, it has been used as a plant fungicide since 1882. It is first synthetic chemical repellent. Its repellent action was investigated in 1935. It was well known repellent for mosquitoes in second world war.

Dimethyl Pthalate

It is liquid in state and was used for general pest control during the second world war. It is effective against *Aedes* and *Anopheles* mosquitoes.

Indalone

It is brown liquid. It was available for use in second world war against mosquitoes as they were responsible for spreading malaria among the civilians and soldiers.

Citronella Oil

It was very commonly used against mosquitoes during the years 1901 to 1938. It is still a popular mosquito repellent.

Rutger-612 or 2-ethyl-1, 3-hexanediol

It is used against mosquitoes and other limited number of house hold pests.

Diacetyl Pthalate

It is effective against cattle fleas.

N, N-diethyl-m-tolumide

It was synthesized in 1952 in USA. It is effective against fleas, deer flies, biting flies, mosquitoes, ticks, etc. *Aedes* mosquitoes are also effectively repelled by this compound.

DIRI-6-2-2 Mixture (Dimethyl) Phthalate 60 per cent by Wt, Indalone 20 per cent, Rutger-612 20 per cent

It was effective against mosquitoes during second world war.

Repellent Colours

Red and blue colours are most effective repellents for mosquitoes.

Physical Repellents

Physical repellents are of following types:

Auditory Repellents

Natural or natural enemy mimic sound is used for repelling insect pests. Amplified sound repels mosquitoes, certain dipteran flies and pyralid moths. Recorded voice or vibrations of natural enemy control the pests like cotton boll worm moths. When sound is produced the moths get repelled as an electronic scare crow.

Contact Stimuli Repellents

The texture of plants produce repellent effect on the tactile sense of the insects. On the texture of plant dusts, granules, oils, waxes, spines, leaf hairs, water, etc. are located which shows bad effect on insect pests by repelling them from the plant.

Visual Repellents

Yellow and blue colours are most repulsive for insects while, white colour is normally attractive to insect pests. However, white is much less attractive than black in mosquitoes and light shades of red or blue attract less than half as many mosquitoes as the corresponding dark shades. Thus, repellent colours and shades do exist and this behaviour can manage pest control. Mosquitoes are guided to their hosts partly visually and partly through sensing temperature gradients and partly by evaporation of water.

Excitatory Repellents

These are substances which excite the insects tarsi by stimulating the sensory nerves and constrain them leaving the treated surface. Such substances are located as insecticides, ex. BHC, DDT, pyrethrum, aldrin, dieldrin, chlordane, lindane, etc.

Barrier Repellents

Several substances act as barrier repellents to insect pests. Mosquitoes nets, wire meshing on doors and windows, plastic enveloping over woollen clothes, tar oil painting, bands on trees and hedges, sleepery bands of tin or plastic to tree trunks, etc are good examples of effective barrier repellents for insect pests.

Feeding Repellents

There are varieties of feeding repellent substances which inhibit feeding in insect pests. Ex. Baygon is effective against cotton boll weevil, Triphentin acetate is effective against cutworms, grasshoppers and several other caterpillars in agro ecosystem. Pyrethrum is effective against *Glossina* sp. (Biting fly).

Repellents for Specialized Feeding Group Pests

Repellents for Foliage Feeders

For flea beetles, leaf hoppers, potato psyllids and cucumber beetles Bordeaux mixture is good repellent. Bordeaux mixture was first used as a repellent in 1882 and it is still in use. For Japanese beetle (*Popillia japonica*) Tetramethyl thiuram disulphide is an effective repellent. A new repellent, 4-(dimethyl triazeno) acetanilide has a wide range of pests. It is effective against cotton leaf worm (*Alabama argillacea*), cotton boll worm (*Anthomus grandis*), the spotted cucumber beetle *(Diabrotica undecimpunctata howardi*) and cabbage looper (*Trichoplusia ni*) while, 6-MBOA is effective against European corn borer (*Ostrinia nubilalis*).

Repellents for Crowing Insects

Creosote, 4,6-dinitro-o-creosol acts as effective repellent for chinch bug. Creosote is good barrier for migratory pests. Termites can be repelled from buildings by treating with trichlorobenzene. Bichlorides of murcury coated on tape can repel ants.

Repellents for Wood Feeders

The wood feeder insects pests like termites, powdered post beetles (*Lyctes* sp., *Anobium* sp., etc.) and Dipterous fly *Pinus* sp., etc effectively repelled from their objects by treating the surface with Pentachlorophenol or Pinosylvian monomethyl ether. Sodium salt also has good insect repellent action. 0.5 per cent aldrin and dieldrin, 0.8 per cent lindane, 1.0 per cent chlordane or 8 per cent DDT in oil solution or water emulsion acts as potential repellents for termites for protecting ground timbers and wooden material of buildings.

Repellents for Fabric Eating Insect Pests

Clothes moths, *Tinea pellinella* and *Tineola bisselliella* and carpet beetles, *Attagenus piceus* and *Anthrenus* spp. are important household pests and cause considerable damage in India. In USA its cost of damage is detected to be $200 millions. Hence moth proofing is usual practice in manufacture of fabrics in USA. Along with dye certain repellent chemicals are incorporated to fabrics. The repellent chemicals used in the process are nontoxic, odourless, colourless, nonvolatile, stable to light, wash resistant and non-damaging. The most widely used repellents in the fabrication are 1-3 per cent "Eulan CN" and "Mitin FF". Wooden garments can be protected by treating them with Naphthalene or moth balls ($C_{10}H_8$). In dry cleaning, oxalic acid 0.03 per cent, sodium fluosilicate 0.6 per cent, sodium aluminium fluosilicate 0.5 to 0.7 per cent water solution, etc are used as preventive and curative measures for clothes moths.

Repellents for Blood Sucking Insects

For blood sucking pests such as mosquitoes (*Aedes* spp. & *Culex* spp.) black flies (*Simulium* spp.) and biting flies (*Tabanus* sp., *Stomoxys*, etc.), Deet and benzyl may be used as a repellents in outdoor environment while, for indoor, Dimethyl phthalate, diethyltolumide (Deet) and Citronella oil are good repellents. Dimethyl phthalate, 2 methy-1, 3-hexanediol were effectively used in second world war against *Anopheles* mosquitoes for preventing malaria spread among the arm force. In anti-malaria programme, Dibutyl phthalate and Benzyl benzoate were also commonly practiced during the second world war. Deet and benzil are new compounds and have special relevance against mosquitoes, biting flies, black flies, etc for outdoor use to workers or sportsman, etc.

Repellents for Domestic Animal Insect Pests

Stable flies, horn flies, deer flies, mosquitoes, face fly, houseflies, etc are important live stock pests. Many insect pests of live stock are responsible for decreasing the yield of milk in domestic animals (cows, buffalo, goats, etc.) and can be repelled by treating the animals with butoxy poly propylene glycol, 2-hydroxyethyl octyl sulfide, synergized pyrithrines, etc.

Repellents for Honey Bees

Bee keeping for collecing honey is an important business for earning foreign currency and improving status of health of the country and agricultural crop yield. At the time of harvestation of honey, and for initial disinfection of beekeeping equipments, honey bees need to be drived from the honey comb. Pheonls, propionic acid benzaldehyde, acetic acid, propionic unhydride, etc. are good repellents for honey bees. Smoke is also used as an effective repellent and it is an oldage practice however, it may lead to forest fires.

Oviposition Repellents

The treatment of pine tar oil diphenylamine on ovipositon sites acts as an effective oviposition deterrent for many insect pests. 6-methoxybenzoxazolinone (6-MBOA) is effective against European corn borer.

Mode of Action

The repellents cause acute disturbances to sensory nerve system which lead to uncoordinated behaviour and thus the insects escape from treated area. For a good repellent, it is essential that an odorous substance must be volatile and that its molecules must contact the olfactory end organ presumably through adsorptive contact. Repellents may exert their influence by masquerading as water absorbed on to the insect olfactory end organ might give the illusion of every high humidity. Pigment molecules in the sensory organs are raised from their ground to excited states by normal metabolic action.

The role of repellents in pest management is crushial and needs more attention for future research on specific repellents and their applications. Anyhow, the ideal repellent should have the following features:

1. Should not be toxic to man and other animals.
2. Should be cheap.
3. Their effects should last longer.
4. Should have wide pest range.
5. Should not be influenced by environmental factors.
6. Should have acceptable tests, odour and touch.

Chapter 8

ANTIFEEDANTS

Today's most pressing problem is to avoid the use of pesticides in pest management. Pesticides, no doubt give quick results but lead to many adverse effects in ecosystems and environment. Antifeedants may solve the problem of excess use of conventional pesticides in the pest control. Insects cause damage to agricultural crops, forest trees, garden plants, live stock, stored grains, man and human dwellings, etc. Thus, their annual loss is more than $ 2,754,000,000 in India, 2,025,000,000 in China, 4,096,000,000 in USA and 5,004,000,000 in USSR.

Definition

"Antifeedants are those chemicals which inhibit feeding in insects and insect die slowly of starvation". Antifeedants are chemicals that inhibit feeding in insects which are applied on the foliage (food) without impairing their appetite and gustatory receptors or driving them away from the food. Antifeedants may further primarily acts as phagodeterrents, gustatory repellents, feeding deterrents and rejectants. In all above cases the insect will die slowly due to starvation.

Antifeedants are classified as follows:

ANTIFEEDANTS

- Triazenes
- Organotins
- Carbamates
- Botanical
  - Pyrethrum
  - Margosa
  - Apple factor
  - Solanum alkaloids
  - Others
- Miscellaneous

Triazenes

It is an odourless and tasteless solid. Chemically it is 4-(dimethyl triazeno) acetanilide. It is most widely field tested compound among the triazenes, and its activated form is 24055. It inhibits feeding in chewing insects like beetles, cockroaches, caterpillars, etc.

Organotins

These are organic compounds of tin. Triphenyltin acetate is one of the earliest organotins used as antifeedant in insects. It is effective against foliage feeding insects such as caterpillars (*Agrotis* spp.), grasshoppers, leaf rollers, colarado potato beetles, etc. According to Lange (1962) organotin antifeedants are effective for the control of chewing pests, preferably lepidoptera. Triphenyltin compounds are effective antifeedants for the hairy caterpillars such as *Pericallia ricini*, *Amsacta albistriga* and *Amsacta moorei* (Verma, 1990). It is also effective against bihar hairy caterpillar, *Spilosoma obliqua*. Triphenyltin acetate has proved to be an effective antifeedant against the fifth instar caterpillars of *Amsacta moorei* at concentrations above 0.025% Triphenyltin hydroxide (TTH) and Triphenyltin acetate (TTA) are good antifeedants for *Helicoverpa* (=*Heliothis*) *armigera* (Lepidoptera). According to Chari & Patel (1973) TTH is superior to TTA as antifeedant for *H. armigera*.

Carbamates

Carbamates are famous for their insecticidal activities. However, a sub-lethal application of some of compounds produce antifeedant effects in certain insect pests. The compounds like thiocarbamates, phenylcarbamates, baygon, etc have antifeedant actions. Thiocarbamates and phenylcarbamates are effective against caterpillars and colarado potato beetle while, baygon is effective against cotton boll weevils.

Plant Products (Botanicals)

Several plants have been tested for their antifeedant activity in insects. The role of plant products in pest management has become crucial since the conventional pesticides lead to several serious problems. *Chrysanthemum*, neem, *Bersama*, *Margosa*, *Anona*, *Calotropis*, *Datura*, *Eucalyptus*, etc. shows good antifeedant activity against insect pests.

Purethrum extracted from the flowers of *crysanthemum cinerariaefolium* acts as an antifeedant for bitting fly, *Glossina* sp. The chemical abyssinin from the plant *Bersama abyssinica* exhibits a strong antifeedant activity against the larvae of *Helicoverpa armigera*. Many plant species have feeding inhibitory action against the third instar larvae of *H. armigera* (Jayraj *et al.*, 1986). Among the different plant products, neem product was found to be promising. In all 19 plant extracts have been tested against *H. armigera*. They are listed below. *Andrographis weightiana* (methanol petroleum ether), *Antigonum leptopus* (Methanol), *Artemisa parviflora* (Acetone), *Adiantum reddianum* (Petroleum ether), *Abutilon persicum* (petroleum ether), *Cassia auriculatum* (methanol petroleum ether), *Clistanthus cellinus* (methanol), *Azadirachta indica* (water), *Coccinia grandis* (methanol), *Cyatheca nilgherensis* (Petroleum ether), *Euphorbia hirta* (Petroleum ether), *Hyptis svauclone* (Petroleum ether), *Marattia fraxinea* (Methanol), *Nicandra physaloides* (Hexane), *Pavetta indica* (Methanol Acetone), *Pedalium murex* (Methanol petroleum ether), *Physalis edulis* (Hexane), *Samadero indica* (Petroleum ether), *Tephrosia maxima* (Methanol petroleum ether). Minimum 40.9% (T.m.) and maximum 88.7% (P.e.) inhibition over control have been observed.

Among datura, neem, A.K. plant and *Cromolaena* leaves tested against *Pericallia ricini* (wooly bear), only neem leaf extract at 5% concentration had a high antifeedant activity. When the concentration of leaf of datura was increased to 5%, it was observed that 80.50% leaf area of brinjal was protected from the feeding *Selepa docilis* indicating its high antifeedant activity. Chauhan & Qadri (1989) reported 88% protection of okra fruits against the fruit borer with 1000 ppm of datura extract. Neem plant extract exhibited high antifeedant property against both *P. ricini* and *S. docilis*. Likely, datura showed high activity towards *S. docilis* and medium to *P. ricini*.

Apogonia blanchardi a chafer beetle, a pest of rose in India, exhibited antifeedant activity against neem seed kernel extract (ethanolic extract). 1% concentration ethanolic extract inhibits the feeding of beetles. It is observed that feeding activity of *Oxycanrnus hyalinipennis* or *Prospis julliflora* is decreased due to ether extract.

Both petroleum ether and aqueous leaf extracts of fifteen plants were tested against second instar grubs of *Henosepilachna viginctioctopuntata* (coccinelinae: coleoptera). *Petroleum* ether extracts at 0.5% and 1% and aqueous extracts at 1% and 5% of test plants, *Annona squamosa* Linn; *Argemone mexicana* Linn; *Calotropis gigantea* R. Br. and *Ricinus communis* Linn. were found to be most potent products which gave cent per cent leaf protection from above mentioned epilachna beetle (Murali Krishna *et al.*, 1990).

A triterpenoid feeding deterrent from *Momordica charantia* L. is isolated and tested against pumpkin beetle, *Aulacophora foveicollis* (Lucas). It is found that 2 mg or more triterpenoid glycoside inhibited the red pumpkin beetle feeding. Bitter gourd leaf extract (BGLE) also have been tested against red pumpkin beetle (*A. fovicolis*) and found good feeding inhibitor for this pest. However, there is a need to structurally analyse the various types of triterpenoids in cucurbits that may help us to pin point as to which specific conformational properties in these molecules or their interactions with each other and other chemicals can be related to inhibitory or stimulatory effect.

Products obtained from neem have attracted considerable attention as potential insect control agent. Neem seed kernel (NSK) extracts and pure compounds obtained from NSK have been reported to produce diverse biological effects on insects such as antifeedant, growth disruptor, oviposition deterrent, toxic, etc of these, antifeedant activity of neem is considered very important. All parts of neem is considered to be biologically active. Methanol extract of seed kernel shows high antifeedant activity as it inhibits feeding at 0.06% concentrations in desert locust, *Schistocerca gregaria* F.

Malus sp. (apple plant) produces phenolic substance called phlorizin which is responsible for probing feeding behaviour in apple aphid (*Aphis poni*) and further acts as a feeding deterrents for the other aphid species like *Myzus persicae* which is not a pest of an apple.

Some plants belongs to the genus *Solanum* produce alkaloids in the form of Leptine, tomatine, solanine, etc. The extracts of these plants shows effective antifeedant activity against potato jassid *Empoasca devastans*. There are many plant substances like non essential aminoacids, tannins and lignans, etc. which acts as antifeedant for insect pests.

Miscellaneous Compounds

Miscellaneous antifeedants include phosphon (2,4-dichlorobenzyl fributyl ammonium chloride), copper resinate, mercuric chloride, copper stearate, etc. which shows antifeedant activity in several insect pests.

Mode of Action

Antifeedants inhibits gustatory receptors of mouth region due to which the insect fails to recognize the treated leaf as food.

Advantages and Disadvantages

Advantages

1. Antifeedants interacts with pests species only (phytophagus) thus, their natural enemies get escaped. Likely, they are not detrimental to pollinating insects and honey bee, etc.
2. Antifeedants are not phytotoxic.
3. Cause no pollution.

Disadvantages

1. Effective against chewing types of insect pests only.
2. Does not cause immediate death of the pest as they may cause damage to crop by moving on untreated plants.
3. Antifeedants alone can not solve the problem of pest management.
4. New growths of plant are not protected.

Chapter 9

NEUROPEPTIDES AND INSECT CONTROL

In pest management, neuropeptides are emerging as promising and new source for controlling insects. Insect neuropeptides are chemical messengers in the insect nervous system. They are primary endocrine regulators. Insect neuropeptides have a role in control of variety of physiological processes like lipid metabolism, carbohydrate metabolism, muscle contraction, metamorphosis, reproduction, growth, water balance, mating, etc. These above physiological processes can be affected through neuropeptides by creating hormonal imbalance that will lead to eco-friendly control of insects. According to Brown (1967) the pentapeptide proctolin was the first neuropeptide identified in insects. The neuropeptides may have direct role in pest control or through other control measures like biological pest control.

The hypothesis, "Neurons, the cellular components of the insect nervous system, produce substances that can regulate and control many biochemical and physiological processes" was first formulated by Kopec in 1917 (1922) and confirmed by Scharrer in 1928. Now the hypothesis is accepted by scientific community that these chemical messengers are oligopeptides or small protein molecules. Proctolin was first detected as a myotropic substance extracted from the hindgut of the cockroach *Periplaneta americana*. According to Starrett and Brown (1975) this peptide is sequenced as Arg-Tyr-Leu-Pro-Thr. Later few other peptides having regulatory role in different physiological processes were identified from insects. Some important neuropeptides identified from the insects are recorded in table 12.

Table 12: Neuropeptides Identified in Insects

Sr.No.	Insect species	Neuropeptide	Reference
1.	*Locusta migratoria*	Adipokinetic hormones AKH-I and AKH-II	Mayer and Candy, 1969
2.	*Schistocerca gregaria*	AKH-I & AKH-II	Beenakkers, 1969
3.	*Periplaneta americana*	Hyperglycaemic hormones HGH-I and HGH-II	Siegert & Mordue, 1986
4.	*Acheta domesticus*	Diuretic peptide (DP-I)	Goldsworthy, personal communication
5.	*A. domesticus*	Achetakinin I-V	-
6.	*Leucophaea maderae*	12-Leucophea peptides	Nachman *et al.*, 1986

Proctolin is detected at least in six insect orders, however, order lepidoptera does not show proctolin. Proctolin causes contraction of insect skeletal muscle and induce myogenic rhythm of contraction and relaxation (Pick & Mantel, 1977). According to Millar (1979) proctolin also has action on the insect heart. Walker *et al.* (1980) recorded action of proctolin on neurons in the insect central nervous system. However, according to O'shea (1986) proctolin is initially considered to be involved in the process of skeletal neuromuscular transmission in a locust, *Schistocera*. Proctolin is gut peptide and its actions are diverse, not restricted to the activity that permits its isolation (Sukumar, 1992).

Master *et al.* (1986) characterized prothoracicotropic hormone (PTTH) as neuropeptide in gypsy moth. Adipokinetic hormone is another important group of neuropeptids. Mayer and Candy (1969) detected a peptide from corpus cardiacum of the desert locust, *Schistocera gregaria* called Adipokinetic hormone (AKH). This hormone is supposed to be responsible for the release of diglycerides from the fat body soon after a flight. AKH interacts with flight muscle and suppress carbohydrate oxidation but stimulates fatty acid oxidation. It is involved in contraction of skeletal muscle and extensor tibialis. At high concentration AKH has a heart accelerating effect. AKH hormone isolated from the corpora cardiaca of variety of insects are found to regulate the metabolism of lipids and sugars. According to Fernlund and Jossefsson (1972) Adipokinetic hormones (AKH-family) are structurally related to the red pigment concentrating hormone (RPCH) derived from the *Pandalus* sp. (Shrimp).

More than 40 insect peptides have been identified, characterized and sequenced of which some important peptides are given in Table 13.

DP-I affect gut contractions in the cockroach *Leucophea maderae*. Cardioacceleratory peptide, FMRF amide (Phe-Met-Arg-PheNH_2) is effective against the insect, *Microcallista nimbosa*. It was first isolated by Price and Greenberg in 1977 from above creature. Evans *et al.* (1989) identified the FMRF amide like peptide by immunoreactive staining. Thus, it is possible to detect the presence of FMRF amide like peptide in a variety of pests. Dockray *et al.* (1981) reported neuropeptides from vertebrates and invertebrates including insects. Holman *et al.* (1986) recorded neuropeptides in a cockroach *L. maderae*. Myers and Evans (1987) demonstrated peptides in *S. gregaria* and Schneider and Taghert (1988) recorded in a dipterous insect *Drosophila*.

Amines like octopamine regulates the secretion of hormones in insect. Chemicals which mimic the activity of such compounds will be good substituents for bringing about hormonal imbalance. Formamidine insecticides are good examples. Development of hormone analogues binds to the receptor but fails to convey any message into the cell and thus fails to evoke any response or hyperstimulate the response. Metabolism of hormones has also received growing attention to manipulate the process in pest control. Degradative enzymes present in haemolymph and other tissue will be the tools of control of pests by manipulating the titer of these enzymes. Various other aspects of neuropeptides such as their synthesis, release and action have also been given attention recently with a view to formulate new strategies for insect pest management. Keeley and Hayes (1987) represented the review of pest management with respect to hormones and neuropeptides.

The discovery of PBAN in *Heliothis zea* provided opportunities for strategies of pest control by interference with behaviour and reproduction by disrupting neuroregulatory events controlling pheromone production (Raina & Menn, 1987). The biosynthesis of neurohormone PBAN can be interrupted by inhibiting the processing enzymes or via synthetic peptidomimetic compounds. Recombinant

Table 13: Primary amino acid sequences of known insect neuropeptides

Name	*Amino acid sequence*										
	1	*2*	*3*	*4*	*5*	*6*	*7*	*8*	*9*	*10*	*11*
Heliothis AKH	pGlu-	Leu-	Thr-	Phe-	Thr-	Ser-	Ser-	Trp-	$GlyNH_2$	-	-
Manduca AKH	pGlu-	Leu-	Thr-	Phe-	Thr-	Ser-	Ser-	Trp-	$GlyNH_2$	-	-
Locust AKH-I	pGlu-	Leu-	Asn-	Phe-	Thr-	Pro-	Asn-	Trp-	Gly	$ThrNH_2$	-
Locussta AKH-II	pGlu-	Leu-	Asn-	Phe-	Ser-	Ala-	Gly-	$TrpNH_2$	-	-	-
Schistocera AKH-II	pGlu-	Leu-	Asn-	Phe-	Ser-	Thr-	Gly-	$TrpNH_2$	-	-	-
Periplaneta M-I	pGlu-	Val-	Asn-	Phe-	Ser-	Pro-	Asn-	$ThrNH_2$	-	-	-
Periplaneta M-II	pGlu-	Leu-	Thr-	Phe-	Thr-	Pro-	Asn-	$TrpNH_2$	-	-	-
Gryllus AKH	pGlu-	Val-	Asn-	Phe-	Ser-	Thr-	Gly-	$ThrNH_2$	-	-	-
Proctolin	Arg-	Tyr-	Leu-	Pro-	Thr-	Ser-	Trp-	$GlyNH_2$-	-	-	-
Leukokinin-I	Asp-	Pro-	Ala-	Phe-	Asn-	Ser-	Trp-	$GlyNH_2$-	-	-	-
Leukokinin-VIII	Gly-	Aln-	Ser-	Phe-	Tyr-	Ser-	Trp-	$GlyNH_2$-	-	-	-
Leucosulfakinin-I	Glu-	Gln-	Phe-	Glu-	Asp-	TyrS-	Gly-	His-	Met-	Arg-	$PheNH_2$
Leucopyrokinin	pGlu-	The-	Ser-	Phe-	Thr-	Pro-	Arg-	$LeuNH_2$-	-	-	-
Achetakinin-I	Ser-	Gly-	Ala-	Asp-	Phe-	Tyr-	Pro-	Trp-	$GlyNH_2$-	-	-
Achetakinin-V	-	-	-	Ala-	Phe-	His-	Ser-	Trp-	$GlyNH_2$-	-	-

DNA technology by cloning genes, coding for neuropeptides and inserting and expressing them in host plants or through microbial agents would worth in insect pest management. Recombinant viruses control insects by behaving as microbial pesticides and by disrupting normal neuropeptide function.

As per the opinion of Menn and Borkovec (1989), a neuropeptide gene placed behind a strong non-essential viral promoter is capable of turning an infected cell into a neurofactory within the insect if the neuropeptide gene retains stability in the virus vector and if cellular processing and transport will mimic the activities of the neuroendocrine tissue. Viruses naturally make a wide hole to the digestive tract, enter into the blood and replicate in host cells from which secreted neuropeptides could be released directly into the blood. The early instar larvae could be prevented from entering the damaging stage and later on could be suspended in development and killed before pupation through neurohormonal deregulation and will lead to biochemical control which is hazard free and pollution free method of pest control.

Neuropeptide control of insects is fit for use in integrated pest management programmes since neuropeptides are highly selective in their action and are active at extremely low concentrations and can be easily engineered by gene technology and delivered directly to the insect through vector system. The use of peptide toxins with specific activity is possible since certain toxins derived from arthropod venoms disrupt feeding or use of food in addition to their acute toxic effects. Anyhow, the use of neuropeptides in pest management is novel control method and good alternative for chemical control.

Chapter 10

FUNGICIDES

The word fungicide has the origin from latin word, *fungus* and *caedo* (caedo means to kill). Fungicides are extensively used in industry, agriculture, houses and garden for protecting crops and their products or protecting carpet and fabrics in houses and the health of human and live stocks.

Definition

Fungicides are those chemicals which have the ability to prevent damage caused by fungi to growing crops and their products.

Fungicides are formulated as wettable and water dispersible powders, dusts and suspension or slurries. They can be applied to seeds, soil and crops. Ideal fungicide should have the characters such as:

1. High field performance
2. Low phytotoxicity
3. Low toxicity to humans and cattle
4. Stability in storage
5. Should posses stability after dilution to spray strength

Fungicide performance is affected by physical factors and chemical factors.

Many recent commercial fungicides are protectants or surface fungicides. They do not penetrate the plant surface or cuticle and are not translocated within the plant but more recent ones can do. Bordeaux mixture is a good example of protectant fungicide.

Fungicides are variously classified. They can be classified on the basis of their: (1) Chemistry, (2) Use, and (3) Mode of action.

CLASSIFICATION OF FUNGICIDES (BASED ON CHEMISTRY)

Sulphur		**Copper**	**Mercurial**		**Ouinone**	**Phenolic**	**Heterocyclic**	**Oxathin**	**Others**
Inorganic e.g.	*Organic e.g.*	*e.g.*	*Inorganic e.g.*	*Organic e.g.*	*e.g.*	*e.g.*	*Nitrogenous e.g.*	*e.g.*	
Powdered sulphur, Wettable sulphur, Lime-sulphur	Ferbam, Ziram, Thiram, Zineb, Maneb, Nabam, Vapam	Bordeaux mixture, Burgundy mixture, Fixed basic coppers like copper oxychloride, Cuprous oxide	Mercuric and Mercurous Chlorides	Phenyl mercury acetate, Methoxy ethyl mercury chloride, etc.	Chloranil Dichlone	Compounds like Dinitrosec-butylphenol, etc.	Captan Folpet	DMOC DCMOD	PCNB, Dodine, Tin Compounds, Dupont 1991, Antibiotics, etc.

Table 14: Fungicide Classifications

Sr.No.	Based on Chemistry	Based on Use	Based on Mode of Action
1.	*Mercurial* *e.g.* Methoxy ethyl	*Seed protectants* *e.g.* Captan, Thiram	*Systemic* *e.g.* Bavistin
2.	*Sulphur* *e.g.* Zineb, Thiram, Sultaf, etc.	*Eradicants* *e.g.* Lime sulphur, Dodine, etc.	*Non systemic* *e.g.* Dinocap
3.	*Copper* *e.g.* Bordeaux mixture	*Blossom protectants* *e.g.* Ferbam, Captan, etc.	—
4.	*Quinone* *e.g.* Dichlone	*Foliage protectants* *e.g.* Captan, Zineb, etc.	
5.	*Nitrogenous* *e.g.* Captan	*Soil fungicides* *e.g.* Dithane, D-14, Captan, etc.	
6.	*Oxathion* *e.g.* Carboxin	*Wound dressing* *e.g.* Bordeaux mixture	
7.	*Benzimidazoles* *e.g.* Benomyl		
8.	*Miscellaneous* *e.g.* Dodine, Antibiotics, natural fungicides.		

Systemic Fungicides (Benzimidazole)

Bavistin ($C_9H_9N_3O_2$)

It is systemic fungicide compatible with various other fungicides, insecticides and acaricides. It is non-phytotoxic and formulated as wettable powder. Bavistin is whitish grey powder, soluble in acetic acid and dimethyl formamide. It is not toxic to bees but, quite effective

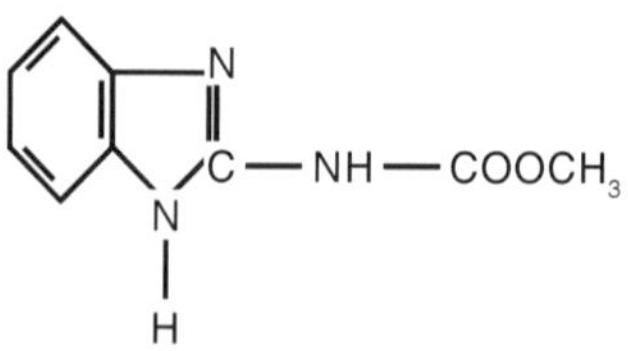

Carbandazim

against ascomycetes, fungi imperfecti and various bisidiomycetes. It is effective against powdery mildew, leaf spot, red rust, stem canker, etc. of several agricultural crops including mulberry. It is also referred as "Carbandazim". Chemically it is methyl-2-benzimidazole carbamate. It translocated within the plant and has both prophylactic and curative actions.

Benomyl ($C_{14}H_{18}N_4O_3$)

It is marked under the trade name "Benlate-50 WP". It is white crystaline, non volatile substance of faint acrid odour. It is insoluble in water and oils. This systemic fungicide is effective against large number of foliar diseases of several agricultural crops including mulberry. It is quite useful for controlling leaf spot of mulberry and other crops. Benlate is chemically [methyl-N-1-(butylcarbamoyl)2-benzimidazole carbamate]. It has systemic, curative and residual activity. It controls rice blast, powdery mildew of cucurbits, legumes and cereals, apple scab, tikka leaf spot of groundnut, etc.

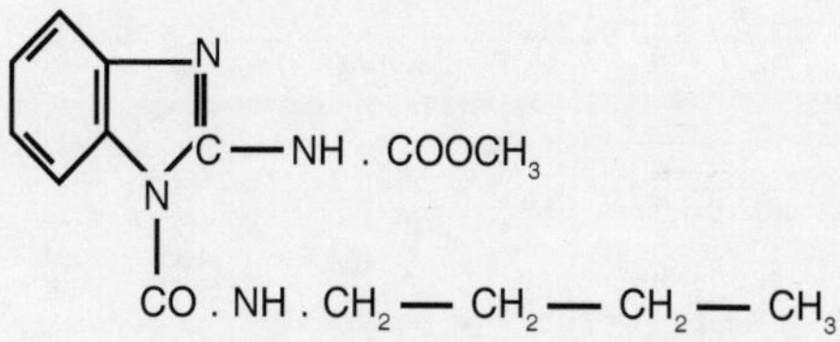

Benomyl

Nitrophenol

Karathane (Dinocap 25% WP) ($C_{18}H_{24}N_2O_6$)

It was introduced in the market in 1946 as dark brown liquid. It is insoluble in water and soluble in most solvents. Chemically it is Dinitro(1-methylhepty) phenyl crotonate (capryl dinitro phenyl crotonate). It is also marketed under different trade names such as Arathane, Iscothan, Mildex, etc. It is effective against powdery mildew of cucurbits and apple varieties. It acts as protectant and eradicant fungicide. It has also acaricidal action. Dinocap is also responsible for removing powdery mildew of rose, tobacco, mulberry and leaf spot of brinjal.

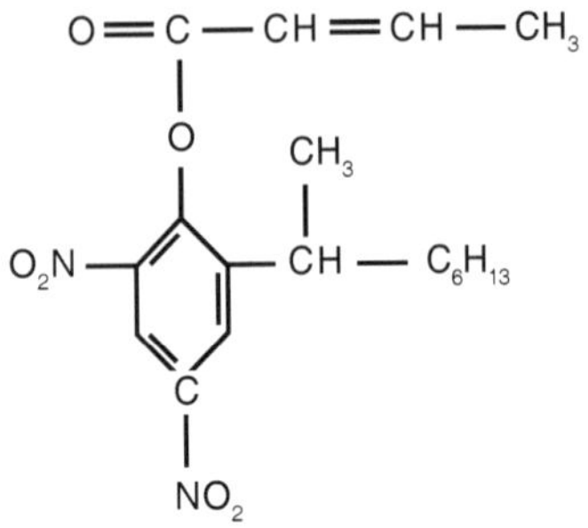

Heterocyclic Nitrogen Compounds

Foltaf 80W (captafol 80% WP)

It was introduced in market in 1976. Now, it is available under different trade names like "Difosan", "Difolaton", "Sanspor", etc. It is effective against leaf spot, leaf rust of several agricultural crops including mulberry. It is also effective against some other diseases (fungal) of agricultural crops.

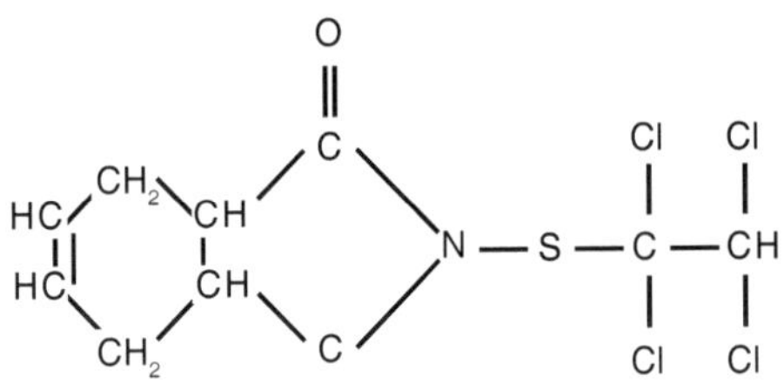

Captafol

Morestan (Chinomethionat)($C_{10}H_6N_2OS_2$)

It is slightly soluble in cold organic solvents but, insoluble in water. It is available as 25% WP and chemically it is 6-methyl-qunoxaline-dithiol-cyclic carbonate. It is very good acaricide and insecticide but, toxic to bees. It is effective against powdery mildew of certain fruit trees, cucurbits, ornamental plants and mulberry. It has low mammalian toxicity but this aromatic nitrogen heterocyclic acaricide has very good control potential against spider mites.

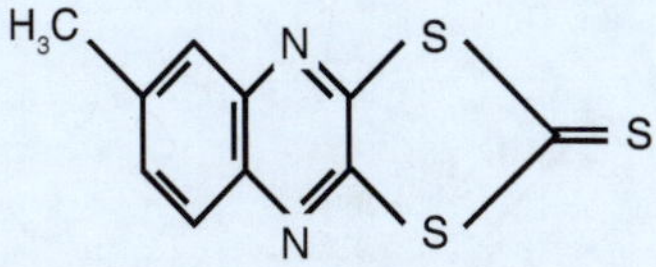

Morestan

Mancozeb (Dithane M-45)(Organic Sulphur)

It was introduced first in 1961 and now available under different trade names such as Dithane M-22, Manzate MEB, and MnEBD. It is effective against twig blight of mulberry, leaf spot and rust of several agricultural crops and many other diseases of agricultural crops (Nene & Thapliyal, 1993). It is also effective tool for controlling muscardine disease of *Bombyx mori*. Manozeb is combine product (Coordination) of zinc ion and maneb. It is available as dust, wettable and liquid flowable powders. Its commercial products available in the market refers to Manzate-200, Manicozin, Manzin, Dithane M-45, Ziman-Dithane, and so on.

Heterocyclic Nitrogenous Fungicides

Captan

Chemically it is N-trichloromethyl thiotetrahydro-phthalimide. Captan is known as Kittleson's killer and orthocide, which is useful against a variety of fungal diseases of fruits, vegetables and ornamental plants. It acts by competing with co-carboxylase for sites on the co-enzyme-free carboxylase in the decarboxylation of pyruvate. As it reacts with SH_2 group, it inhibits many oxidative enzymes. Captan is incompatible with all alkaline materials. Hence, incompatible with Bordeaux mixture and lime-sulphur, but it is compatible with other common fungicides. It is comparatively safer fungicide. Captan, a persistent foliage fungicide effectively controls apple and pear scab, black spot on rose and powdery mildew of several agricultural and horticultural crops. It is also responsible for supressing *Gloeosporium* rot of apples, stem rot of tomatoes and *Botrytis* mould on soft fruits. It is easily synthesized from the readily available starting materials butadiene and maleic anhydride.

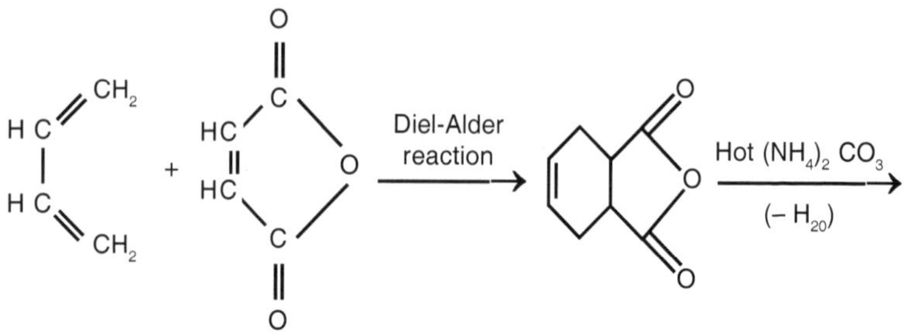

Butadiene **Maleic anhydride**

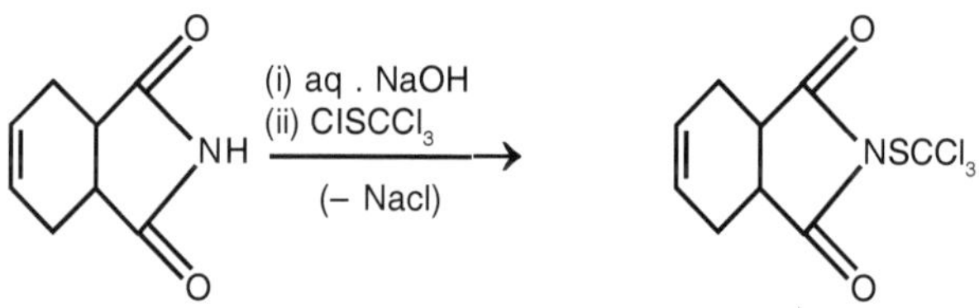

Tetrahydrophthailimide **Captan**

Trichloromethyl sulpphenyl chloride can be collected from carbon disulphide.

$$CS_2 + 3\,Cl_2 \xrightarrow{I2} Cl_3\,C.SCl + SCl_2$$

It controls apple scab, bitter rot, black rot, black pox, fly speck, blossom end rot, etc with 0.2% spray. It also controls leaf spot of banana with 0.2% spray and citrus leaf spot, brown rot; coffee brown eye spot; grape grey mould, downy mildew and black rot; jute stem rot; maize seed rot, seedling blights; onion smut, pea damping-off; peach brown rot, leaf curl, bacterial spot; potato fungal seed, early blight, late blight; strawberry leaf blight, fruit rot, leaf spot; tomato stem rot, early blight, damping off, late blight, anthracnose, leaf spot, grey leaf spot and wheat burnts.

Folpet

It is analogue of captan which was first introduced in california for controlling apple diseases. Folpet is referred as "Phaltan" and found more effective than captan. Captan has movable hydrogens to carbonyls but, this is lacking in folpet. Luken & Horsfall (1964) found that folpet imide is 100 times more toxic than the captan imide to spores of *Monilinia* and *Stemphylium*. Folpet imide melts at 100°C. Chemically folpet is N-(trichloromethylthio) phthalimide. It controls apple sooty blotch, fly speck, bitter rot, black rot and scab with 0.1 to 0.15% spray. It also controls rose black spot and tobacco brown spot with the dose 0.2% spray.

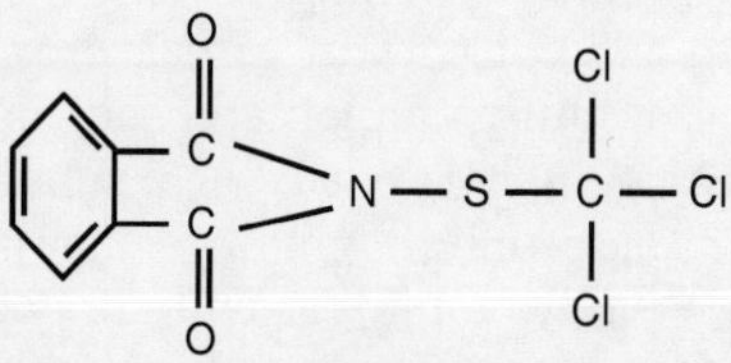

Folpet

Quinone Fungicides

Chloranil

Chloranil was first introduced in 1940 in USA under the trade name "spergon". It is good seed dressing fungicide. It is golden yellow with molecular weight 245.9. It is toxic to man. In light and moisture it is changed into chloranilic acid. It is insoluble in water but completely soluble in ether. It is compatible with most commonly used seed dressing insecticides. It is effective against downy mildew

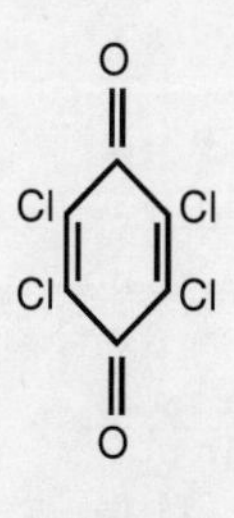

Chloranil

of tobacco, onion and cabbage. It controls Barley loose smut; daming off of beans, cinchona, coffee, etc; maize seed decays, seedling blight; pea root rot, seed decays; sorghum grain smut; sugarcane pineapple disease; sweet potato stem rot, black rot and wheat bunts, etc. Chemically it is 2,3,5,6-tetrachloro-1, 4-benzoquinone.

Dichlone

Chemically it is 2,3-dichloro-1, 4-naptho-quinone. It is released in the market under different trade names like "phygon", "phygon XL", etc. Its pure form is odourless yellow crystal. It is insoluble in water but partly soluble in xylol, benzene and acetone. It is stable in sunlight and has molecular weight of 227. It is seed protectant fungicide for rice, jowar, pepper, tomato, peas, etc. It also acts as foliage fungicide against certain diseases of agricultural or horticultural crops. It controls apple scab; beans anthracnose, peach leaf curl; pea damping off; potato red xylem, tuber decay; sorghum grain smut, sugar beet seedling blight; sugarcane pineapple disease and tomato leaf spot, grey mould, etc. with spray dose 0.1 to 0.3%.

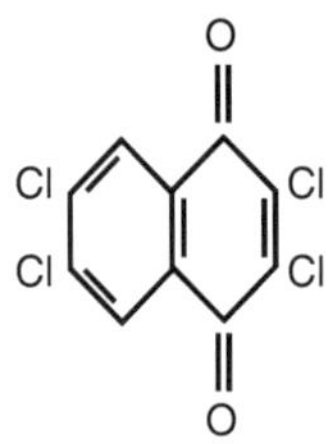

Dichlone

Sulphur Fungicides

Lime-sulphur

Lime sulphur has been used for more than 80 years for protection of fungal diseases on agricultural and non-agricultural crops in India. Lime sulphur is a mixture of rock lime, sulphur and water with various proportions. It is mainly available as dormant spray. A proportion of lime sulphur is given below :-

Rock sulphur	-	20 lbs.
Sulphur	-	15 lbs.
Water	-	50 gallons.

The standard form of lime sulphur has baume density of 31-33 degrees at 60°F and contain less than 20% calcium polysulphides. It acts as contact fungicide and effective against different types of diseases (fungal). It is more effective against powdery mildews of crops. It controls apple scab, powdery mildew; cherry leaf spot; plum black knot; rose powdery mildew; pear black spot; peach leaf curl, powdery mildew, brown rot; grape powdery mildew, etc.

Organic Sulphur (Carbamate) Fungicides

Ziram

Ziram is marketed under different trade names such as "Corozate", "Fuklasin", "Karban white", "Milbam", "Vancide SIZ", "Zerlate" and "Zirberk". Chemically it is zinc dimethyl-dithiocarbamate. It is white substance with molecular weight 305.8.Ziram is soluble in chloroform and water (at 65 ppm). It is stable fungicide and toxic to human and other mammalians. It is useful against number of diseases of vegetable crops and ornamental plants. It is available for use as spray. It controls apricot frecke, brown rot; apple bitter rot, scab, bull's eye rot; anthracnose of tobacco, bean, cucurbits, dioscoria, tomato; mint rust; peach curl; ragi blast; sugarbeet leaf spot; beans rust; potato early blight, etc.

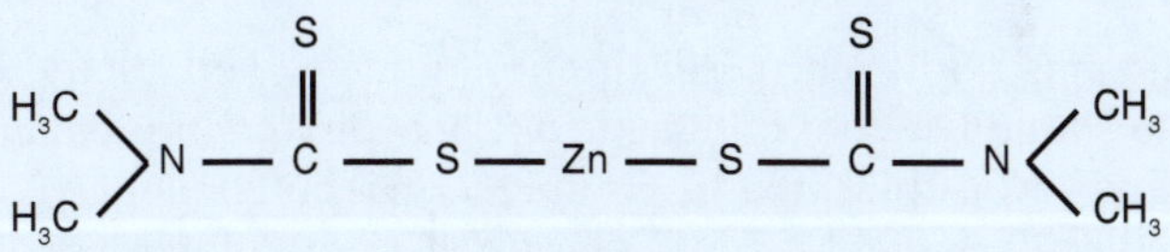

Ziram

Thiram

It is marketed under different trade names such as Arsan, Panoram[75], Pomasol, Tersan[75], Thylate, TMTD, etc. Thiram-75 is common form in India. Chemically it is tetramethyl thiuram disulphide or bis(dimethylthiocarbamoyl) disulphide. It is white coloured substance with molecular weight 240.4. It is completely soluble in chloroform and acetone but insoluble in water. It is unstable in presence of acids but compatible with most commonly used pesticides except few (copper & lime component). It is effective seed protectant. As like ziram, it cause irritation to nose, throat and skin. It is effective against several fungal diseases of crops. It controls damping off disease of flax, brinjal, cabbage, turnip, beans, pea, onion, spinach, tobacco and sugarbeet; seedling diseases of citrus, radish, okra; bitter rot, fruit rot, fly speck, cedar rust, black pox of apple; rust, brown rot and freckle of apricot; seed rot anthracnose of beans; chickpea blight; coriender stem gall; cabbage club rot; cucurbits downey mildew, wilt, stem blister, groundnut crown rot, seed decay, grey mould; jute stem rot; maize black stem, seed decay and seedling blight, stalks rots; pea smut, root rots, stem rot; peach brown rot, leaf curl; rice poor germination, sheath rot, leaf blight; safflower rust; strawberry leaf spots, fruit rots; sugar beet rust, black leg; sunhemp stem break; tobacco and sweet potato scurf, stem rot, wheat root rots, etc.

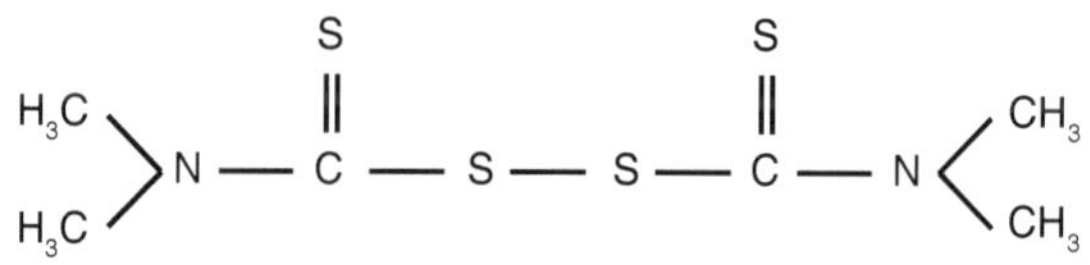

Thiram

Zineb

Zineb is marketed under different names like Dithane Z-78, Parzate C, and Du Pont Fungicide A, Lohacol. Chemically zineb is zinc ethylene bis-thiocarbadte. Its pure form is buff or light coloured with molecular weight 275.5. It is partly (10 ppm) soluble in water. It is changed to other form in presence of light, heat or moisture. It is

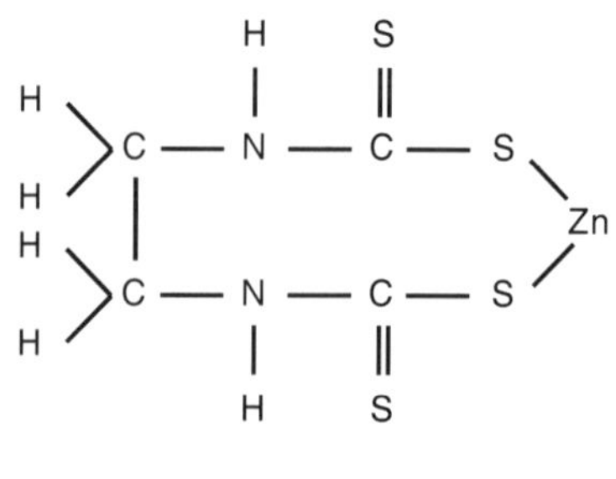

Zineb

compatible with most of the common fungicides and insecticides. Zineb is toxic to man which cause irritation to skin, nose and throat. It is effective against a very large number of crop diseases. It is useful as foliar sprays on vegetables and fruit trees to control leaf spot, downy mildew, blight, scab and anthracnose diseases. It can be used as sprays with 0.15 to 0.3% and also as dusts.

Maneb

It is marketed under different trade names such as Dithane M-22, MEB, MnEBd, etc. Chemically it is manganese ethylene bisdithiocarbamate. Maneb is yellow coloured and slightly soluble in water. It has molecular weight 265.3. It is effective against a very large number of fungal diseases of agricultural and horticultural crops.

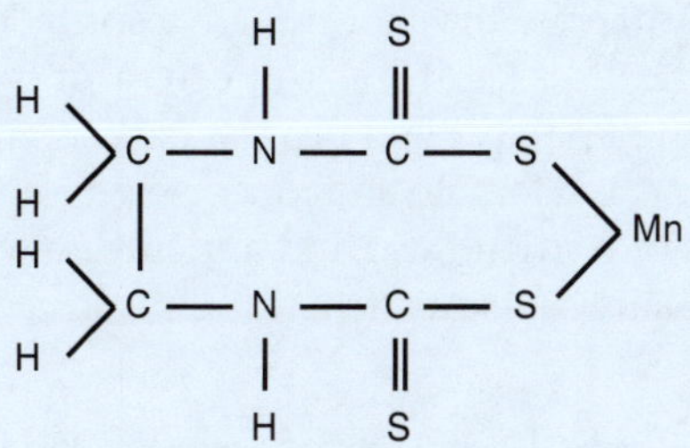

Maneb

Nabam

It is marketed under different trade names like Chembam, Dithane A-40, Dithane D-14, Parzate liquid, etc. Chemically it is ethylene bisdithiocarbamate. It has molecular weight 256.3. It is liquid in form. But the powder form is unstable in presence of heat, light and moisture. It is effective as foliar spray and for soil application and seed treatment of cotton. At 0.2% spray it is effective against barley leaf disease, beans rust, carrot leaf spot, maize leaf blights; onion downy mildew, blast, leaf blast; pepper ripe rot, sweet potato stem rot; tomato leaf mould, grey leaf spot; wheat stem rust, leaf rust; peturia leaf blight, etc.

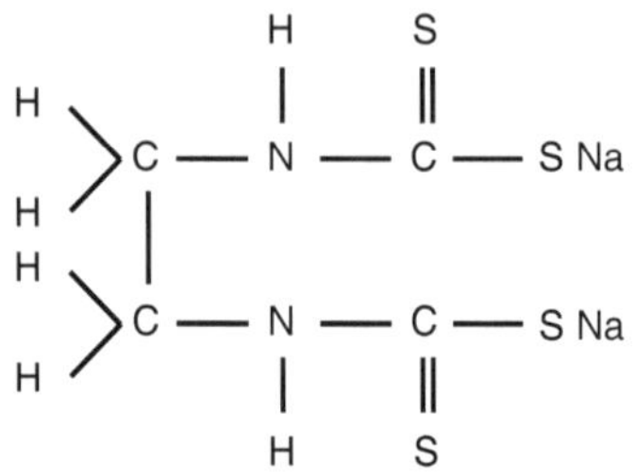

Nabam

SDMC (Vapam)

It is marketed at VPM, Chemvepe, 4-S, etc. Chemically it is sodium methyldithiocarbamate. It is soluble in water it has BP of 119°C. It can be applied with watering can or sprinklers. It effectively controls soil borne diseases. It is also effective for controlling cotton wilt, damping off of papaya and peas, sugarbeet root rot, seedling diseases of tobacco and tomato and root disease of tea. It also control several diseases of horticultural and agricultural crops caused by nematodes, specially root knot nematodes.

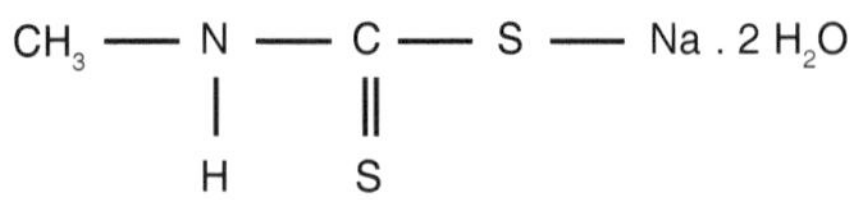

SDMC

Copper Fungicides

Bordeaux Mixture

French scientist, Millardet accidentally observed the fungicidal activity of Bordeaux mixture in 1882 on the crop grape. The mixture of copper sulphate and lime is the "Bordeaux mixture" or "Bouillie Bordelaise". Powdery mildew of grapes was controlled accidentally. Later, in 1887, after conducting successful tests, Millardet exposed BM as good fungicide for use in fungal diseases. In 1888 it was successfully tested against potato late blight. Millardet proposed following proportion for Bordeaux mixture for effective and safe guard results (prevention of germination of spores)-

1. 4 kg of blue stone
2. 4 litre lime
3. 50 litre of water i.e. 4-4-50

Copper sulphate and lime are separately added in sufficient quantities of water. In case of quick lime, slaking is done slowly. These two solutions are then mixed simultaneously into a third container, while agitating thoroughly, and the bordeaux mixture is prepared. Readymade mixture by some European countries is marketed as dry or in paste forms. The inherent solubility of BM is quite low. It is insoluble in water. Freshly prepared bordeaux mixture should be used immediately because, fungitoxicity of mixture decreases on standing. Bordeaux mixture controls downy mildew and powdery mildew of grapes; apple scab, fire blight; cherry bacterial canker; citrus canker; coconut leaf rot; grapes black rot; groundnut leaf spot; maize downy mildew; mango anthracnose; papaya leaf spot; pan leaf spot; okra leaf spot, etc. with different proportion of BM.

Systemic Fungicides

Chloroneb (Demosan)

It was tested in 1823 under the name Fungicide 1823 and marketed under the trade name "Demosan 65W". Chemically it is 1,4-chloro-2, 5-dimethoxy-benzene. It is slightly soluble in water (7 ppm). It is effective against fungal diseases of cotton, beans, peas, cucurbits and peanuts. It is good soil fungicide.

PP 675 (Methyrimol)

Chemically it is 5-Butyl-2-dimethylamino-4-hydroxy-6-methylpyrimidine. It is white crystalline solid with molecular weight

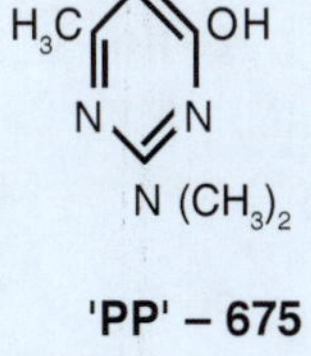

'PP' – 675

209.28. It is soluble in chloroform but very little in water. It is effective against *Sphaerotheca* on cucumbers, melons and squashes, but not so effective against powdery mildew of vines.

Cycloheximide (Actidione)

Chemically it is β-3[2-(3,5-dimethyl-2-oxocyclohexyl)-2-hydroxyethyl] glutarimide. It is marketed under different trade names such as Actispray, Actidion Rz, etc. It is slightly soluble in water. It is effective against several diseases of horticultural and agricultural crops which includes leaf spots, rusts, bunts, powdery mildew, etc.

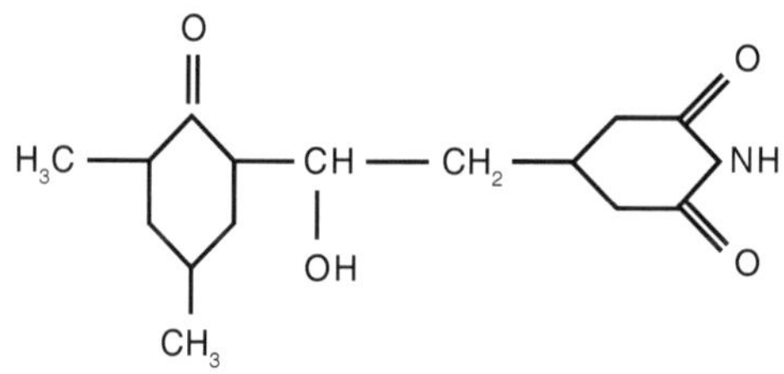

Actidione

Miscellaneous Fungicides

Dodine

It was released in the market under the trade name "dodine". Chemically it is n-dodecyl-guanidine acetate. It is crystalline solid with molecular weight 287.3 and M.P. 136°C. It is insoluble in most of the solvents but soluble in hot water and alcohol. In moderate acid and alkaline condition dodine can be stand as stable compound. It has bad effects with human as it cause eye and skin irritation. It is effective against spot diseases of ornamental plants and horticultural plants. It can control apple scab, capsicum bacterial leaf spot, loquat black spot, rose black spot, strawberry leaf spot, pear scab, etc.

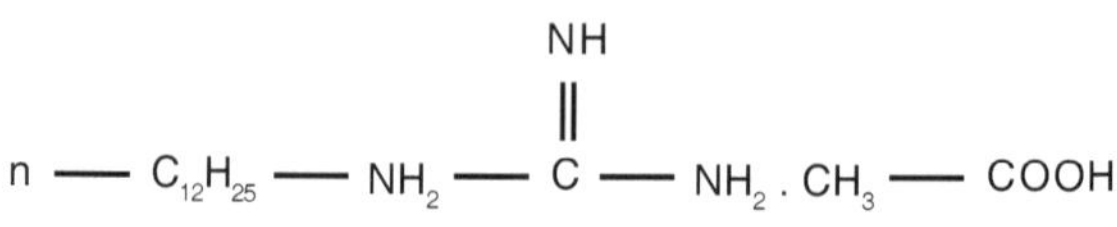

Dodine

Quintozene (PCNB)

It was investigated in 1930 in Germany. In 1935 PCNB was successfully tested against *Rhizoctonia; Scleroctinia*, etc. in Holland and released in 1956 under the trade name "PCNB" in USA. Chemically it is pentachloro-nitrobenzene. It is faint yellowish brown substance with molecular weight 295.4 and M.P. 140-143°C. It is soluble in toluene and benzene. It may cause skin irritations but less toxic to warm blooded animals compared to dodine. It is available with different trade names such as Terrachlor, Quintozene, Brassicol, Botrilex, Tritisan, etc. and effective against several fungal diseases of plantation crops, agricultural crops and horticultural crops. It controls daming off of soyabean, papaya, crucifers, cotton, coffee, beans, etc. Likely, it also control cotton seedling disease, groundnut root rot, potato scab and black scurf; wheat bunts, sugarbeet sclerotium rot, etc.

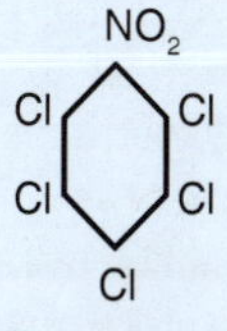

PCNB

TPTA (Duter)

It is marketed under different trade names such as Duter, Brestan, TPTH, etc. It is tin compound and chemically it is Triphenyltin acetate (TPTA). It controls potato late blight, cucumber anthracnose, gummy stem blight, downy mildew, etc.

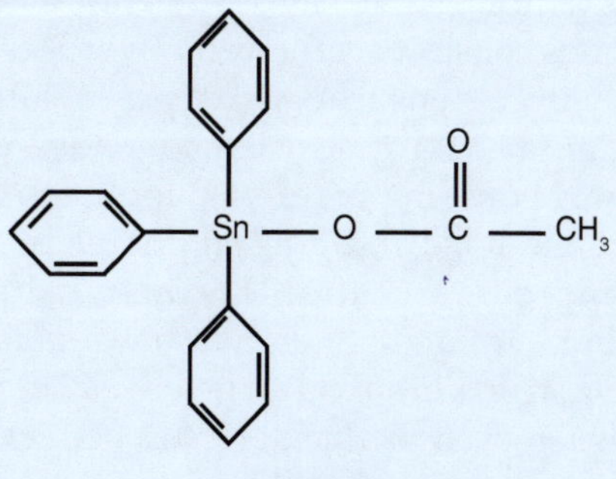

TPTA

Chapter 11

WEEDICIDES

Out of 2,50,000 plant species in the world, weed plant comprises about 250 species. Weeds are those unwanted plants which interfere with crops for land and water resources and affect the yield of the crops. The weeds can cause about 45% damage to agricultural crops and are responsible for reduction in land value, loss of quality, reduction in human efficiency, increase in cost of insect and disease control, etc. Hence weed control is an important task for hoping maximum yield in the crops. The weeds are classified into three groups namely, annual (ex. Tumble mustard) biennials and perenials (ex. red top). Each group is further subdivided as dicotyledonous weeds (broad leaved weeds) and mono cotyledonous weeds (grasses).

Weeds are controlled by weedicides or herbicides. More than 400 herbicides have been investigated during last 55 years. Weed problem can be perfectly solved by weedicides. Herbicides are organic chemicals primarily made up of carbon (C) and hydrogen (H) atoms. The carbon atoms form chains. It may be straight, cyclic or branched. Organic compounds consist of carbon and hydrogen hence called hydrocarbons. Hydrocarbons are of two types (i) saturated and (ii) unsaturated. In saturated form all available bonds are occupied by an atom of C and H while in unsaturated form two carbon atoms share more than one bond resulting in double or triple bonds. Unsaturated ring configuration are also called aromatic hydrocarbons. Organic compounds that have atoms other than 'C' as a part of ring structure are called heterocyclic hydrocarbons. There are many heterocyclic herbicides. Most herbicides contain at least one halogen atom. Alcohols (R-OH), organic acids (R-COOH) and esters (R-COOR) are other forms which affect water solubility, electrical charge and volatility. Hence, they have important role in weed control. Herbicides are available in the form of liquid or solid and are applied at

1. Pre-planting
2. Pre-emergence and
3. Post-emergence in relation to crop and weed and also applied either to the soil or foliar.

Types of Herbicides

CLASSIFICATION

Herbicides					
Action Based		*Chemistry Based*		*Application Based*	
Physio-logical	Biochemical	Organic	Inorganic	Soil applied	Foliage applied
				1. Systemic 2. Contact	1. Systemic 2. Contact

Table 15: Classification of Organic Herbicides Based on Chemistry and Mode of Action

Sr.No.	*Chemical Group*	*Soil Applied*	*Foliage Applied*
1.	*Acetamides*	Acetochlor, alachlor, dimethenamide metolachlor, napropamide, pronamide propachlor	Acetochlor, alachlor dimethenamide metolachlor, propanil
2.	*Alipatics*	–	dalapon, Acrolein*
3.	*Arsenicals*	–	DSMA, MSMA, cacodylic acid*
4.	*Benzamides*	Isoxaben	–
5.	*Benzoics*	Dicomba	–
6.	*Benzothiadizoles*	–	Bentazon*
7.	*Bipyridiliums*	–	diquat*, paraquat*
8.	*Carbamates*	–	Asulam, desmedipham, phenmedipham

Contd...

Table 15–contd...

Sr.No.	*Chemical Group*	*Soil Applied*	*Foliage Applied*
9.	*Cineoles*	Cinmethylin	–
10.	*Cyclohexanediones*	–	clethodium, sethoxydim cycloxidim
11.	*Dinitroanilines*	Benefin, dinitramine, ethalfluralin, isopropalin, nitralin, pendimethalin, oxyzalin, prodiamine, trifluralin	–
12.	*Diphenyl ethers*	Fluorodifen, nitrofen* oxyfluorfen,	bifenox, lactofen, oxyfluorfen, fluoroglycofen acifluorfen
13.	*Imidazolidinones*	Buthidazole	Buthidazole
14.	*Imidazolinones*	imazapyr, imazaquin	imazaquin, imazapyr, imazethapyr
15.	*Imines*	–	CGA-245757
16.	*Isoxazolid inones*	clomazone	–
17.	*Nitriles*	–	bromoxynil*, ioxynil*
18.	*Oxadiazoles*	oxadiazon	–
19.	*Oxodiazolidines*	methazole	methazole
20.	*Phenols*	–	Dinoseb*
21.	*Phenoxy alkanoic acid*	2,4-D	2,4-D, MCPA
22.	*N-phenylphthalimides*	–	flumiclorae
23.	*Phenyl pyridazine*	–	pyridate
24.	*Phthalamates*	naptalam	naptalam
25.	*Pyrazoliums*	–	difenzoquat
26.	*Pyridazinones*	purazon, norflurazon	–
27.	*Pyridine carboxylic acids*	–	picloram, trichorpyr, clopyralid
28.	*Pyridines*	dithiopyr, thiazopyr	–
29.	*Pyridinones*	Fluridone	Fluridone

Contd...

Table 15–contd...

Sr.No.	Chemical Group	Soil Applied	Foliage Applied
30.	*Pyrimidinylthio benzoates*	pyrithiobac	pyrithiobac
31.	*Sulphonylureas*	Chlorsulfuron, chlorimuron, bensulfuron, sulfometuron, halosulfuron	bensulfuron, metasulfuron, nicosulfuron, triasulfuron, etc.
32.	*Thiocarbamates*	Thiobencarb, EPTC, triallate, pebulate, butulate, diallate	–
33.	*Trizines*	Prometon, cyanazine, hexazinone, atrazine, ametryn, etc.	Prometryn, prometron, atrazine, etc.
34.	*Triazinones*	Metribuzin	–
35.	*Uracils*	UCC-C4243 bromacil, terbacil	UCC-C4243
36.	*Ureas*	Diuron, linuron, monoron, siduron, isoproturon, fluometuron, etc.	diuron, linuron, tebuthiuron, etc.
37.	*Unclassified*	Ethofumesate	Fosamine, glyphosate

Several inorganic herbicides have been investigated and utilized in weed control between 1896-1930's. Important among them are given below:-

Copper sulphate, sodium arsenite, ferrous sulphate, copper nitrate, ammonium sulphate, sulphuric acid, sodium chlorate, sodium nitrate, sodium chloride, sodium tetraborate, sodium metaborate, etc. In recent days, these weedicides are useless due to nonspecificity and their less effectiveness. Now a days, organic herbicides are playing very important role in weed control.

Herbicides (Weedicides)

Acetomides

Characters

1. Mostly applied at pre-emergence (PRE), sometimes "POST" actively.
2. Monocots appear twisted and malformed.

3. Leaves tightly rolled and do not emerge properly.
4. Broad leaved weeds show cupped or crinkled leaves.
5. Shows drawstring effect on tip.

Alachlor

Chemically, it is 2-chloro-N-(2,6-diethyl phenyl)-N(methoxymethyl) acetamide. It is the first herbicide developed in 1966 but, marketed under the trade name "LASSO" in 1969. It acts as effective herbicide for the crops jowar, maize, groundnut, cotton, soyabean etc. It is effective against several kinds of weeds like *Panicum* spp., *Setaria* spp., *Digitaria* spp., *Protulacea* spp., *Cyperus esculentus, Echinochloa crus-galli, Solanum nigrum, Eleusine indica*, etc. It is applied at pre-planting incorporated (PPI), pre-emergence (PRE) and early post emergence (E-POST) about 2-4 kg/ha for above crops. The herbicides give good results within 6-10 weeks. Alachlor's field half life is 21 days. Its residue do not persist longer in the environment. Its synthesis involves methoxy methylation of acetanilide. It is metabolised fastly by *Cyperus esculentus* weed. Alachlor is detoxified by conjugation with glutathione (GSH) or homogultathione (hGSH). It affects fatty acid biosynthesis by inhibiting acetylelongase in weeds.

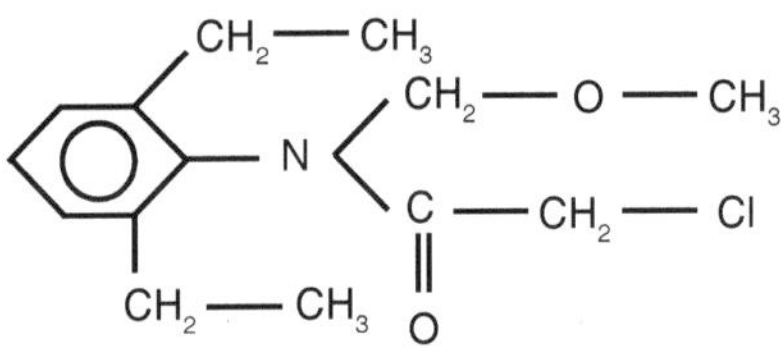

Alachlor

Butachlor

Chemically it is N-(butoxymethyl)-2 chloro-N-(2,6-diethyl phenyl) acetamide and was investigated in 1970 but Mosanto commercialised this under the trade name "MACHTE" in 1971 as useful weedicide. It is world famous herbicide for rice. However, it is effective against several kinds of weeds in crops rice, pulses, potato, soyabean, groundnut, etc. Under irrigated conditions or equivalent rainfall condition, it is applied at PRE. In rice it is recommended to apply wtihin 2-5 days after transplantation of seedlings or after 10-15 days of direct sowing (seed) crops. It controls many annual grass weeds like *Panicum* spp., *Setaria* spp., *Echinochloa crus galli, Digitaria* sp., etc. within 4-7 weeks. Butachlor is available as E.C. (50 EC) (Emulsifiable concentrate) and granules (5G). Granules are applied in combination with 2-4-diethyl ester (4G). Its field half life is 13 days. It is metabolized fastly, primarily to polar water soluble metabolites. It is detoxified by conjugation with GSH and hGSH. It affects fatty acid biosynthesis by inhibiting the enzyme acetylelongase in weeds and also acts as protein synthesis inhibitor.

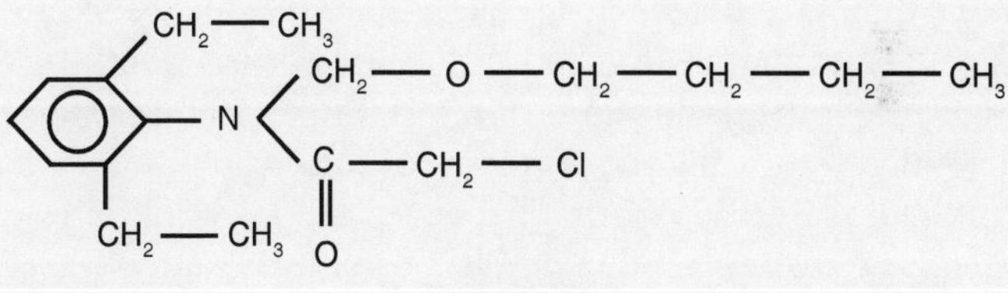

Butachlor

Metolachlor

2-chloro-N-(2-ethyl-6-methyl phenyl)-N-(2-methoxy-1-methyl ethyl). It was synthesized in 1976 by Ciba Geigy and marketed for use against weeds of maize agro-ecosystem in 1977. It is effective herbicide to control weeds in the crops like sorghum, safflower, groundnut, potato and soyabean. It controls *Setaria* spp., *Digitaria* spp., *Brachiaria* spp., *Cyperus esculentus, Panicum capillare, P. dicotomiflorum* and *Echinochloa crus-galli*. It is also effective against broad leaved grass weeds such as *Mullugo verticillata, Amaranthus retroflexus*, etc. It is applied at E.PPS, PPI, PRE and E-POST with a dose of 1.5 to 4.00 kg/ha in maize, PRE at 1.5 to 3.00 kg/ha in jowar,

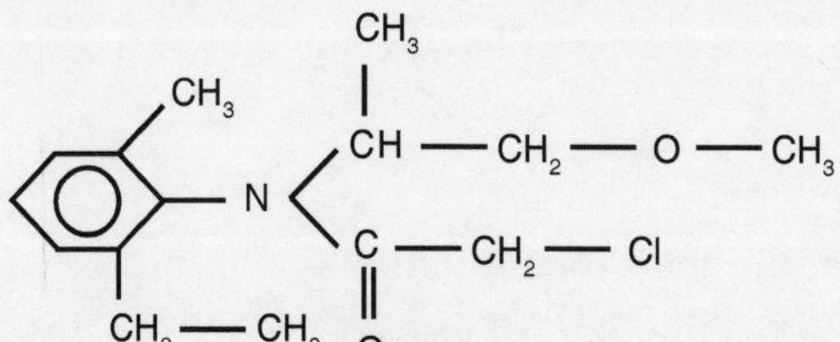

Metolachlor

soyabean and safflower. It is equally effective at PPI, PRE or after hill formation in potato with 1.5 to 3.00 kg/ha.

Metolachlor has its field half life period of 3-5 months, however, residues does not persist longer. It controls the weeds within 10-14 weeks. It can be mixed with attrazine, metrinuzine, cyanazine, etc. During the synthesis of metolachlor, reductive alkylation and subsequent chloracetylation of the alkylamine takes place. It acts as inhibitor of fatty acid (lipid) biosynthesis and protein synthesis.

Dimethenamid

Chemically it is 2-chloro-N[(1-methyl-2-methoxy)ethyl]-N-(2,4-dimethyl-thien-3-yl)-acetamide. Dimethenamid (SAN 582) was investigated and synthesized by Sandoz and released for the use as herbicide under the trade name "FRONTIER" in 1993 for maize. It was also successfully tested for control of soyabean weeds in 1994. It is applied at E-PPS, PPI or PRE about 0.8-1.4 kg/ha in agro-ecosystems of soyabean and maize. It controls many annual grasses like *Panicum dichotomiflorum, Echinochloa crus-galli, Digitaria* spp. and *Setaria* spp. and broad leaved weeds such as *Amaranthus retroflexus, Solanum nigrum*, etc. Its field half life is 20 days. It is used as a mixture with atrazine. It can also be applied with water or liquid fertilizer. Dimethenamid acts as inhibitor of protein synthesis in weeds.

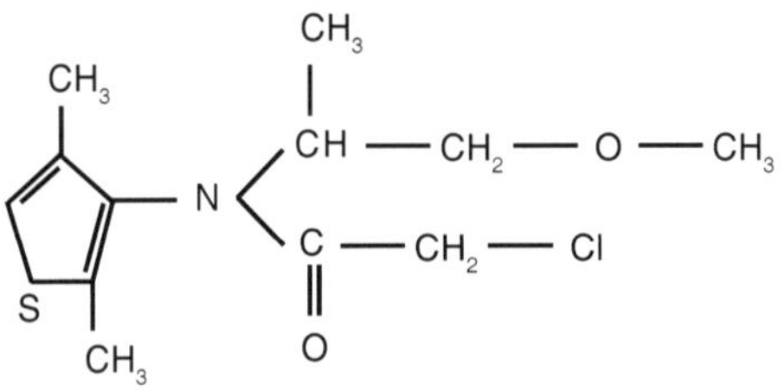

Dimethenamid

Pronamide

Chemically the pronamide is 3,5-dichloro(N-1,1-dimethyl-2-propynyl) benzamide. It was developed by Rohm and Hoas in 1969 and released under the trade name "KERB". It has another name propyzamide. It is effective weedicide for the crops like vegetables, fruits, cotton, sunflower and groundnut. It is applied at PRE with 1.5 to 4.00 kg/ha. It is effective against the weeds like *Digitaria, sanguinalis, Chenopodium album, Portulaca oleracea, E.C. galli* and many more annual broad leaved weeds and grasses. It may have "POST" activity against some grasses. Its field half life is 60 days. It can be synthesized by the reaction of 3,5-dichlorobenzyl chloride with 3-amino-3-methylbutyne. It acts as inhibitor of protein synthesis and lipid biosynthesis in weeds.

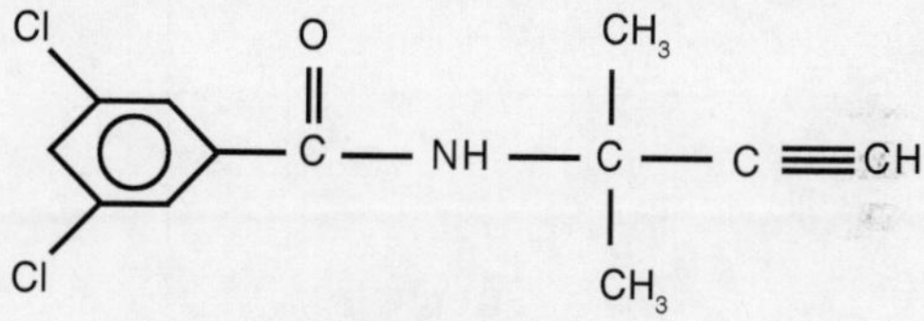

Pronamide

Propanil

Chemically it is 2-chloro-N(1-methylethyl)-N-phenyl acetamide. It is an acetamide herbicide released in 1961. It is a very good herbicide for the crops wheat and rice. It controls the weeds like *Eleusine indica, E. C. galli, Setaria* spp., *Amaranthus retroflexus, Digitaria* spp, etc and several other grasses and broad leaved weeds. It is applied at 'POST' with 3.00 to 5.00 kg/ha in rice and 1.0 to 1.25 kg/ha in wheat. Its field half life is only one day. It is used as a mixture with molinate. Its residues are not harmful to crops planted the following season. It is synthesized by reacting propionic acid and 3,4-dichloroaniline to form N-(3,4-dichloro phenyl) propanamide which reacts with propionic anhydride followed by vacum distillation.

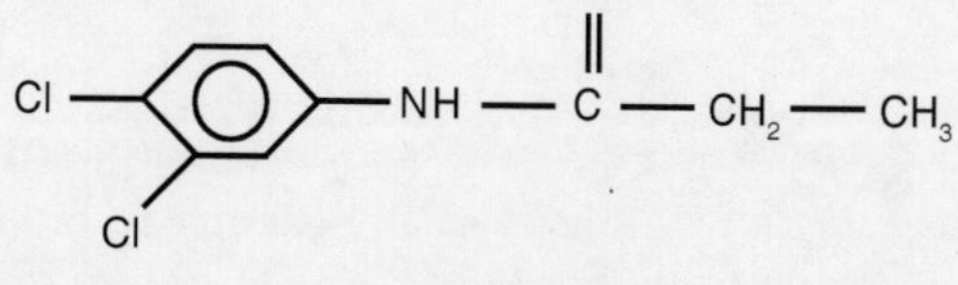

Propanil

Acrolein (2-propenal)

This weedicide is from aliphatic campaign and also recognized by the second name, acrylaldehyde. It was released as weedicide in 1930. It is an aquatic and effective contact weedicide against *Pistia stratiotes*, *Jussiaea* spp. etc. It is hazardous hence great care should be taken during its use. It is also toxic to fishes and wildlife.

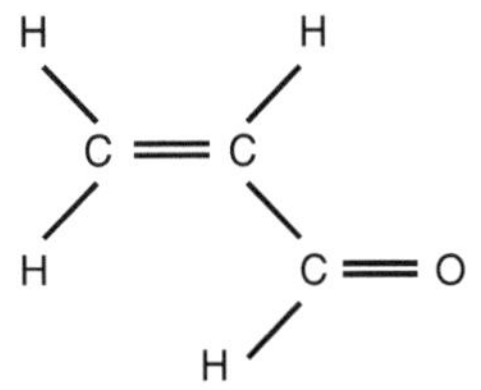

Acrolein

Arsenical Weedicides

Characters

1. Investigated in 1951 but widely used since 1971.
2. Shows low to medium mobility on sandy soil but largely immobile on other soils.
3. Affect by chlorosis and necrosis.
4. Non degradable in soils.
5. Provide effective control to parenial weeds.
6. Examples - DSMA, MSMA, Cacodylic acid etc.

Monosodium Methanearsenate (MSMA)

It is a foliage weedicide and effective in cotton agro-ecosystem to control various weeds at 2.5 kg/ha. It is useful at POST and can be mixed with diuron, floumeturon, 2,4-D dimethylamine, dicamba,

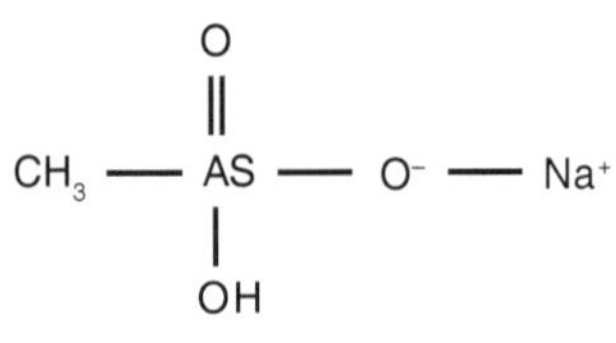

MSMA

mecoprop etc. Its field half life is 180 days and has moderate to long residual activity in the environment. It acts as membrane disrupter in weeds, inhibits oxidative phosphorylation and tissue respiration. It controls following weeds *Setaria* spp., *Digitaria* spp., *Xanthium* spp., *Paspalum dilatatum, E. indica, Amaranthus retroflexus, Ambrosia artemisifolia, Sorghum halepense,* etc. and many annual broad leaved and grass weeds.

Disodium Methane arsenate (DSMA)

DSMA is strongly absorbed by soil. It is effective weedicide in the crops, citrus, cotton and turf. Its field half life is 180 days. When applied POST at 3.5 kg/ha, 2.5 kg/ha and 2.5 to 3 kg/ha in turf, cotton and citrus respectively, it effectively controls the following weeds - *Setaria* spp., *Digitaria* spp., *Amaranthus retroflexus, Ambrosia artemisifolia, Sorghum halepense, Cyperus rotundus, Xanthium strumarium* etc. It acts as membrane disrupter and inhibits oxidative phosphorylation in weeds.

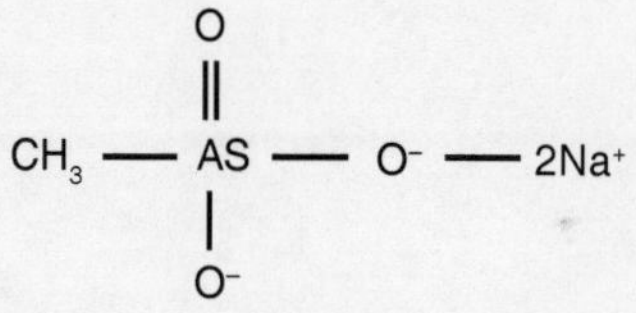

DSMA

Benzoics

The benzoics have been recognized as herbicides since 1958. It was Dicamba who showed herbicidal properties.

Characters

1. Twisting and curling of stems and petioles.
2. Stem swelling at nodes and elongation.
3. Leaf cupping.
4. Chlorosis,
5. Inhibition of growth.
6. Wilting and necrosis.

Dicamba

Chemically Dicamba is 3,6-dichloro-2-methoxy benzoic acid. It is effective against weeds in crops like jowar, maize, wheat, turf, etc. It is available as a water soluble alkali in the form of spray and granules. It is applied PPS, PRE at 0.5 kg/ha in maize while in jowar, wheat, turf and even in maize it is applied as POST at 0.25 kg/ha. It is effective against many broad leaved weeds. Its field half life is not more than 14 days. It can be used as mixture with 2,4-D, atrazine, MCPA, etc. It acts as protein biosynthesis stimulant.

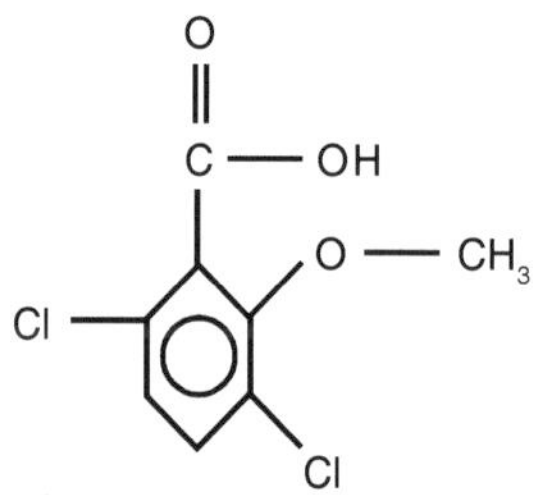

Dicamba

Benzamides

Isoxaben

Chemically it is N-[3-(1-ethyl-1-methyl propyl)-5-isoxazolyl]-2,6-dimethoxy benzamide. It was investigated in 1979 but marketed in 1982. It is effective against weeds in tobacco crop with PPI at 1.5 to 2.5 kg/ha. It controls several annual grasses including *E.c.galli, P.*

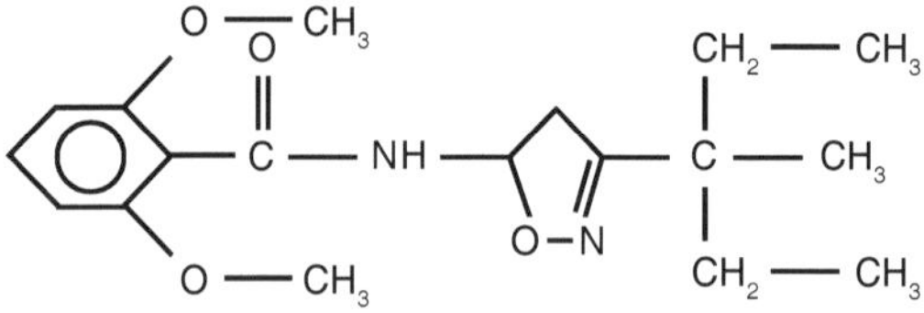

Isoxaben

dichotomiflorum, *Setaria* spp., *Digitaria* spp. and broad leaved weeds like *Mullugo verticillata, A. retroflexus* and *Chenopodium album* etc. Its field half life is 100 days. It can be mixed with ozyzalin and trifluralin. It acts as growth inhibitor in broad leaved weeds.

Benzothiadiazoles

Bentazon

Chemically it is 3-(1-methylethyl)-(1H)-2,1,3-benzothiadiazin - 4(3H)-1,2,2,-dioxide. It was reported in 1968 as "BASAGRAN". It is effective weedicide for controlling the weeds in the crops, jowar, maize, wheat, barley, rice and soyabean. It controls the weeds such as *Cyperus* spp. and the weeds of the families like compositae, cruciferae, cyperaceae, solanoceae, polygonaceae, amaranthaceae, ambrosiaceae, etc. It cause chlorosis in sensitive plants within 3-5 days. Its field half life period is 20 days. It has very less or no residual activity. For the synthesis of this weedicide, anthrinilic acid is reacted with N-isopropylamino sulphonyl chloride and further formation of ring structure with phosgene takes place. It acts as photosynthesis inhibitor.

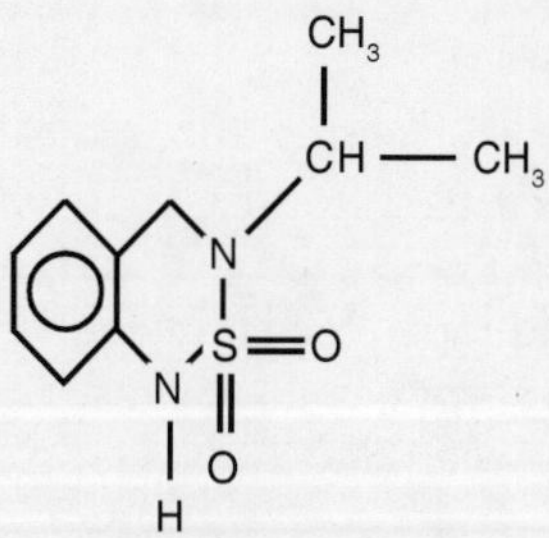

Bentazon

Bipyrilidiums

Characters

1. Synthesized from puridines.
2. Phytotoxic.
3. Contact in action.
4. Available in the form of salts.
5. Widely used herbicide members are Paraquat and Diquat.

Paraquat (1,1'-dimethyl-4,4'-bipyridinium ion)

It was used as growth regulator in 1959. However it is recognized as most widely used herbicide in the world. It is effective weed control agent in several crops like tea, coffee, pineapple, potato, sugarcane and cotton. Its field half life is 1000 days. It can be applied as pre-harvest desiccation of potato and cotton but it is applied POST at 0.25 to 1.00 kg/ha for several weeds. Biologically it is unavailable to microbes and it is completely inactive in soil. It is available in the form of dichloride and di(methyl sulphate) salt. It causes rapid wilting desiccation in weeds and complete necrosis within a few hours of application. Paraquat is synthesized by quaternization of 4,4'-bipyridyl with chloromethane under pressure with or without solvent. It is photosynthesis inhibitor of weeds.

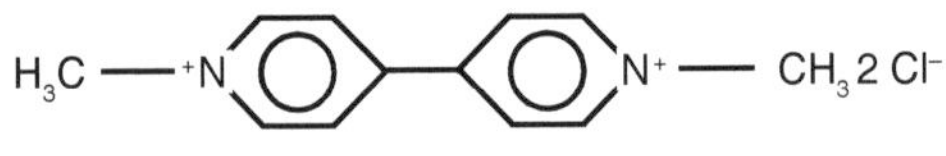

Paraquat

Diquat

It was released as growth regulator in 1955. It is effective weed control agent in the crops cotton, castor, jowar, soyabean, sugarcane, etc. It is applied as POST spray to wet, at 0.2 to 0.4 kg/ha for controlling many aquatic weeds. Its field half life is 1000 days. It affects photosynthesis of weeds. It causes necrosis within 1-3 days to weeds. It controls weeds like *Salvinia* spp.,*Centella* spp., *Eichhornia* spp., etc.

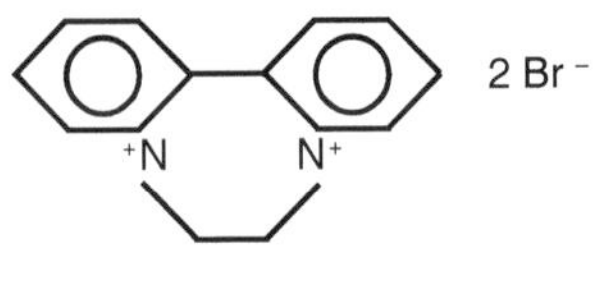

Diquat

Carbamates

Characters

1. These are esters of carbanilic acid.

2. First discovered in 1945.
3. Carbamates are diverse group of herbicides.
4. Effective weedicide members includes asulam, phenmedipham, desmedipham, etc.

Asulam (Methyl-1[4-aminophenyl sulfonyl]carbamate)

May and Baker investigated its herbicidal properties in 1965 as a sodium salt. It is a good weedicide for weed control in sugarcane plantation and orchard crops. At POST application with 1.00 to 3.00 kg/ha it controls several grass weeds such as *Sorghum halepense, Eleusine indica, Digitaria sanguinalis* and perennials like *Pteridium aqualinum* (bracken fern). Its field half life is 7 days. It causes chlorosis in young leaves and stunting and necrosis in weeds. It also acts as DHP inhibitor.

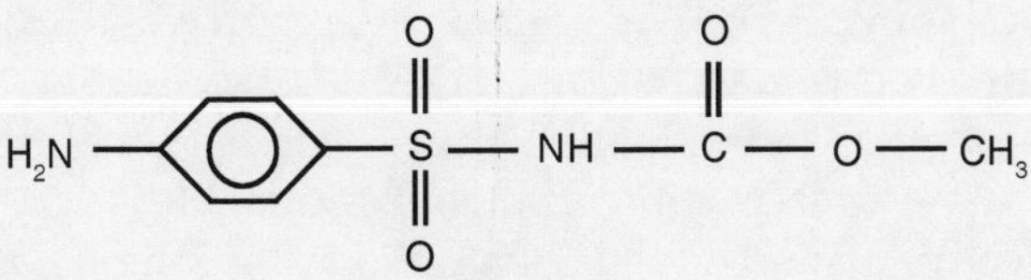

Asulam

Phenmedipham

Chemically it is 3[(methoxy carbonyl)amino]phenyl(3-methyl phenyl) carbamate. Its weedicidal activities were tested in 1966 and released for the use in 1967 in Europe. It controls weeds such as *Setaria* spp. and annual broad leaved weeds like *Portulaca oleracea* and *Solanum nigrum* at POST with 0.4 to 0.7 kg/ha in the crops like sugarbeets and vegetables. Its field half life is commonly 44 days but varies with climatic conditions. It causes damage to weeds by water soaked appearance chlorosis and necrosis.

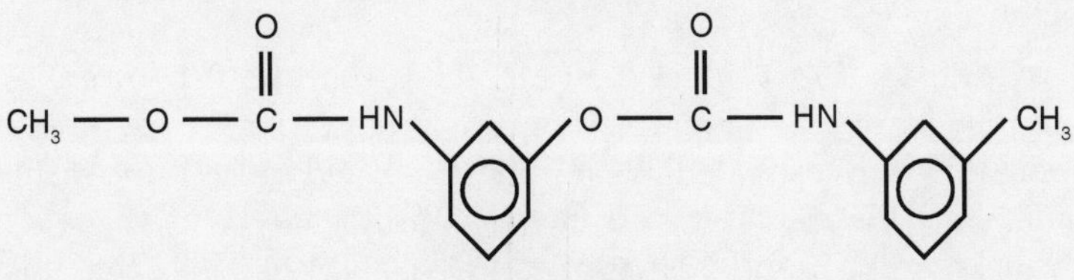

Phenmedipham

Cyclohexanediones (Cyclohexenones)

Characters

1. Acts as growth inhibitor.
2. Young leaves are affected first.
3. Causes brown chlorosis.
4. Before necrosis leaves become purple, orange and red.
5. Acts as root growth inhibitor.
6. Discovered first in 1980.
7. Effective against annual and perennial grasses.
8. Important examples include Sethoxydium, clethodim etc.

Cycloxidim

Chemically it is 2-[1-(ethoxyimino)butyl]-3-hydroxy-5-(2H-tetrahydrothiopyran-3-yl)-2-cyclohexen-1-one. It was discovered in 1984 by BASF. It is marketed as "FOCUS" and "LASER" for the control of annual and perennial grasses. Its field half life varies with environmental conditions but, its residual life is very short. It is not effective against broad leaved weeds. It inhibits fatty acid (lipid) biosynthesis in weeds.

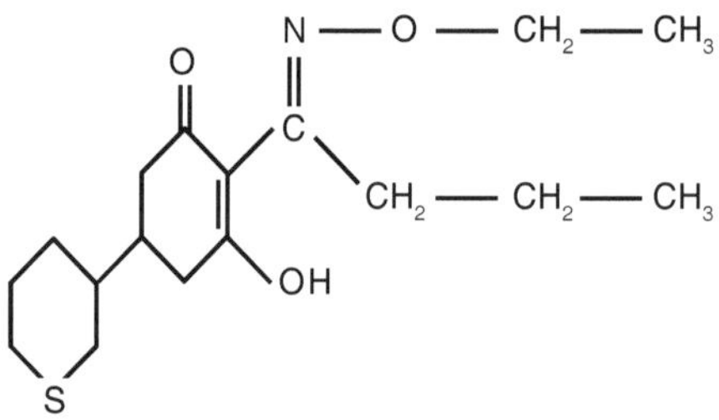

Cycloxidim

Clethodim

Chemically it is (E,E)-(±)-2-[1-(3-chloro-2-propenyl) oxylimino)propyl]-5-[2-(ethylthio)propyl]-3-hydroxy-2-cyclohexen-1-one. It was investigated in 1987 and marketed in 1991 under the trade name "SELECT". It is effective against the weeds in crops like soyabean and cotton when applied POST at 0.1 to 0.3 kg/ha. It is not effective against broad leaved weeds but useful to control several annual and perennial grasses. It is non persistent in soil. It has field half life of 3 days. The compounds like bentazon and acifluorfen inhibits the activity of clethodim. As like cycloxidim it is inhibitor of fatty acid (lipid) biosynthesis in weeds.

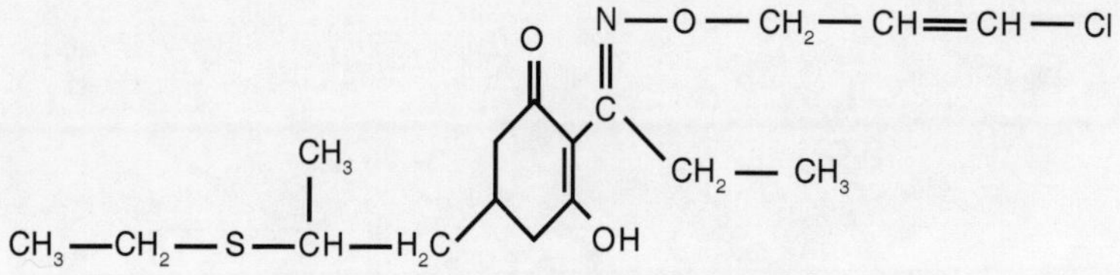

Clethodim

Tralkoxydim

It is effective weedicide for annual and perennial grasses at POST with 0.1 to 0.4 kg/ha in tea, coffee, plantation crops, fruit crops, wheat, soyabean, cotton, etc. Its field half life is 3-4 days.

Dinitroanilines

Characters

1. Orange yellow coloured compounds.
2. Less water soluble.
3. Volatile in nature.
4. Strongly absorbed by soil.
5. Acts as root growth inhibitor.
6. Root tips become thick.

Trifluralin

Chemically it is 2,6-dinitro-N,N-dipropyl-4-(trifluromethyl) benzenamine. Its herbicidal properties have been reported first in 1960 and have been widely used as soil herbicide in 1964. It is a very famous member of Dinitroanilines campaign. It is effective herbicide in the crops such as safflower, castor, sunflower, pulse crops, vegetable crops, maize, wheat, cotton, soyabean and horticultural crops. It can be applied at PPI with 0.56 to 1.12 kg/ha. It controls many annual grasses and broad leaved small seeded weeds. It can be applied through irrigation system or impregnation on dry fertilizers in sorghum wheat and soyabean at 0.84 to 2.24 kg/ha. As a 'POST' with 0.84 to 1.40 kg/ha it can't be applied to sugarbeets, onion and potato. Its residue lasts longer in the environment. However, its field half life is just 45 days in moist soil. It is the inhibitor of microtubule assembly in weeds.

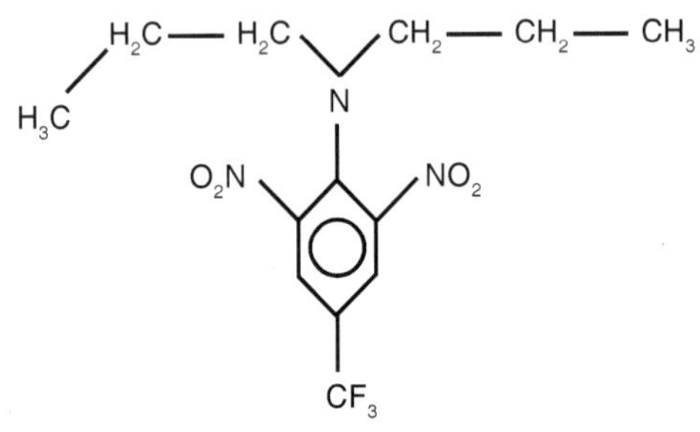

Trifluralin

Ethal fluralin

Chemically it is N-ethyl-N-(2-methyl-2-propenyl)-2,6-dinitro-4-(trifluoromethyl). It was released as herbicide in 1974. It is effective against the weeds in the crops such as sunflower, groundnut, soyabean and vegetables at PPI with 0.6 to 1.2 kg/ha. It is also effective to control weeds in wheat agro-ecosystem. It controls many annual grasses like *Setaria* spp., *Digitaria* spp., *Ponicum*

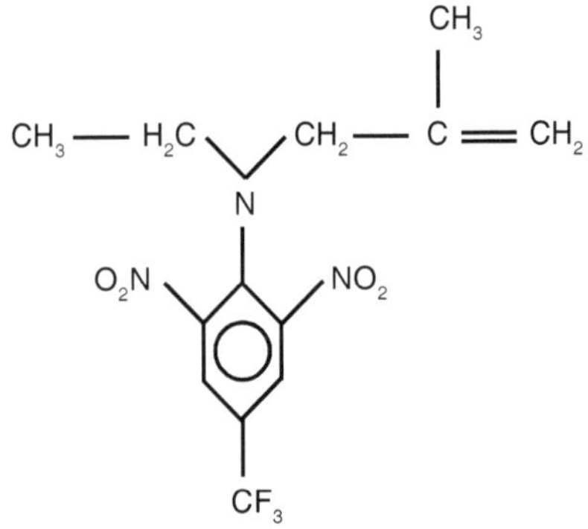

Ethalfluralin

dichotomiflorum, E. c.galli, etc. and broad leaved weeds like *Kochia scoparia, Solanum nigrum, Amaranthus retroflexus,* etc. It is phytotoxic to certain sensitive crops even after one year of application. It inhibits microtubule assembly as like trifluralin.

Fluchloralin

It is chemically N-(2-chloroethyl)2,6-dinitro-N-propyl-4-(trifluoromethyl)aniline. It is an effective weedicide for control of several annual broad leaved weeds and grasses when applied as PRE in crops such as vegetables, pulses, cereals and oilseeds like groundnut, etc.

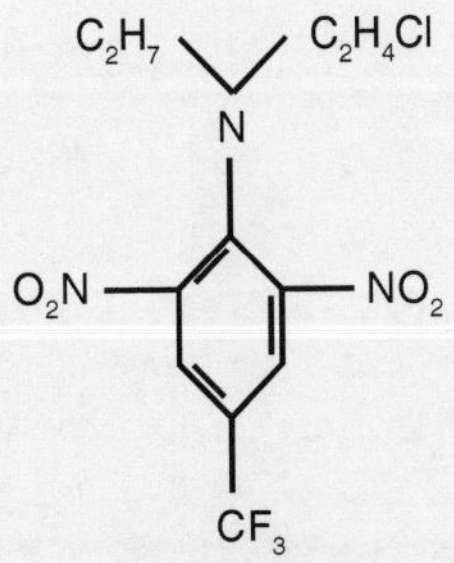

Fluchloralin

Benefin

Its weedicidal properties have been reported first in 1965. It controls annual grasses like *E. c.galli, Setaria* spp., *Digitaria* spp., *Brachiaria* spp., *Panicum* spp., etc. in groundnut, turf and tobacco

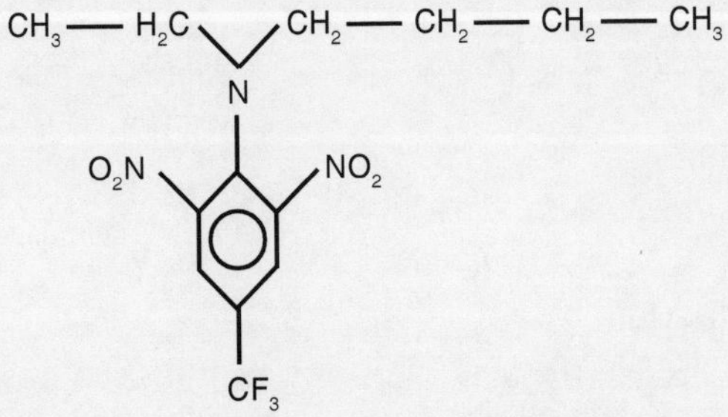

Benefin

when applied PPI at 1.25 to 1.60 kg/ha. It also controls some annual broad leaved weeds. Its field half life is 40 days.

Diphenylethers

Characters

1. Contact in action.
2. Not well translocated in plants.
3. Within 1-2 days chlorosis and necrosis formed.
4. Applied both PRE and POST.
5. Examples include bifenox, lactofen, formesafen, acifluorfen, oxyfluorfen, etc.

Acifluorfen

It is chemically 5-(2-chloro-4-(trifluoromethyl)phenoxy]-2-nitrobenzoic acid. It is released as weedicide by American Cyanamid under the trader name "SCEPTER". It is effective against the crops groundnut and soyabean at 0.15 to 0.40 kg/ha at POST. It is also effective in rice at 0.15 kg/ha. Acifluorfen controls several annual broad leaved weeds such as *Datura stramonium, Amaranthus* spp., *Polygonum* spp., *Brassica kaber, Solanum nigrum, Ambrosia artemisiifolia, Sesbania exaltata, Xanthium* spp., etc. and grass weeds like *Sorghum helpense, Setaria* spp., etc. Its field half life is about 14-60 days depending on environment. Its residues does not persist in the environment for longer period.

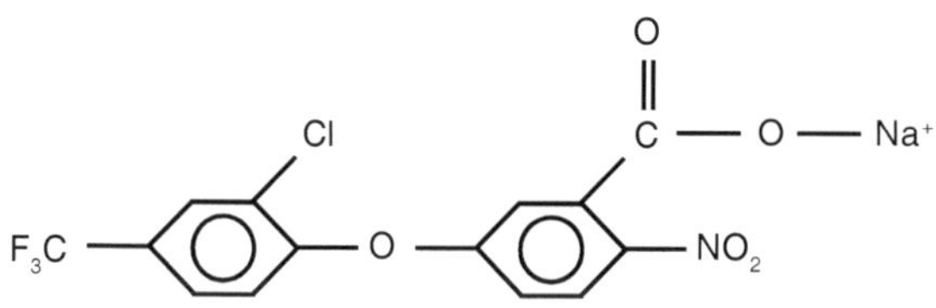

Acifluorfen

Oxyfluorfen

Chemically it is 2-chloro-1-(3-ethoxy-4-nitrophenoxy)-4-(trifluoromethyl)benzene. It was registered as weedicide under the trade name "GOAL" in 1976 but, its weedicide properties have been investigated in 1975. It is effective against the weeds in plantation crops such as rubber, tea, coffee, coconut, palms and leguminous crops at 0.25 to 2.00 kg/ha. It is also effective weedicide in groundnut, soyabean and rice. As a POST directed spray it is effective weedicide in the crops such as cotton, maize, soyabean, fruit trees and plantation crops. It controls several annual grasses including *E. c.galli, Digitaria* spp., *E. indica, Paspalum conjugatum, Axonopus compressus* and broad leaved weeds like *Amaranthus* spp., *Borreria hispida, Commelina benghalenesis, Portulaca oleracea*, etc. Its field half life is 35 days and moderately persistent by residues. It is synthesized by reacting resorcinol with 3,4-dichlorobenzotrifluoride. The resultant is then nitrated and finally ethoxylated. It acts as pigment biosynthesis inhibitor.

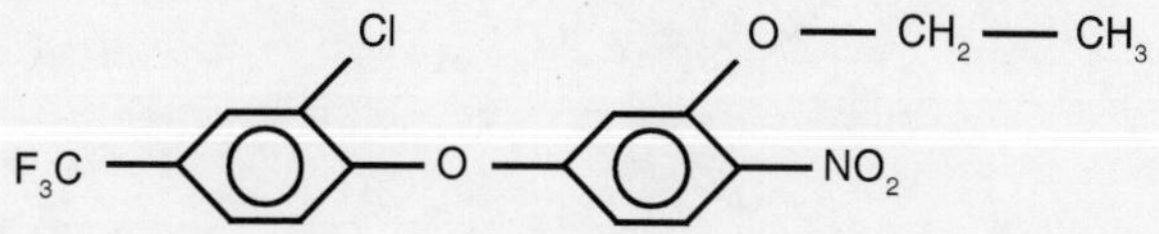

Oxyfluorfen

Imidazolinones

Characters

1. Acts as growth inhibitor in weeds.
2. Meristematic area becomes chlorotic.
3. Cause foliar chlorosis and necrosis.
4. Examples include Imazethapyr, Imazaquin, Imazapyr, Imazamethamenz, etc.

Imazaquin

Chemically it is 2-[4-5-dihydro-4-methyl-4-(1-methyl ethyl)-5-oxo-1H-imidazol-2-yl]-3-quinoline carboxylic acid. It was investigated in 1980 but its weedicidal properties were reported in 1981 and marketed for use under the trade name "IMAGE" and "SCECTER". Commercially it is available as ammonium salt. It is widely used as weedicide to control the following weeds - *Datura stramonium, Kochia scoparia, Amaranthus* spp., *Ambrosia* spp.,

Euphorbia spp., *Xanthium strumarium*, *Impomoea* spp., *Chenopodium album*, *Solanum nigrum*, *Anoda cristata*, etc. as broad leaved annual weeds. Imazaquin also controls many annual grasses including *Digitaria* spp., *Setaria* spp., *E.c.galli*, *P. dichotomiflorum*, *Sorghum halepense*, *S. bicolor*, etc. It is applied PRE or POST at 50-70 g/ha in maize, groundnut, pulses and soyabean. Imazaquin is synthesized by dehydration condensation and hydrolysis of 2,3-quinoline-dicarboxylic acid to form anhydride. It then forms a ring structure called imidazolinone ring and gets precipitated.

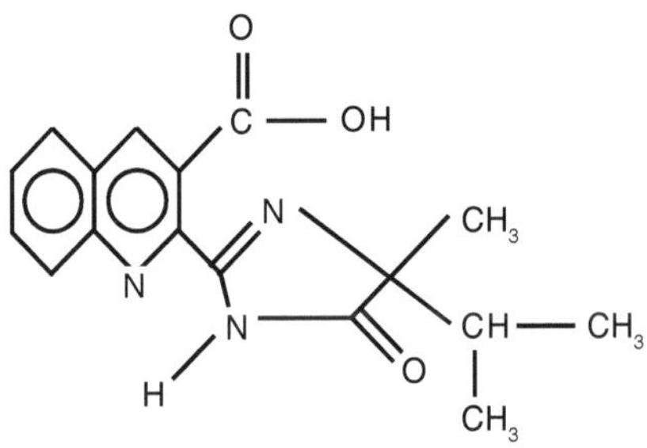

Imazaguin

Nitriles

Bromoxynil

Chemically it is 3,5-dibromo-4-hydrobenzonitrile. It was investigated in Germany in 1896 and its herbicidal properties have been demonstrated in 1963. It effectively controls the broad leaved weeds like *Abutilion theophrasti*, *Brassica kaber*, *Capsella bursa pastoris*, *Solanum* spp., *Polygonum convolvulus*, *Chenopodium album*, *Chorispora tenella*, etc. It is applied at POST with 0.21 to 0.56 kg/ha in crops

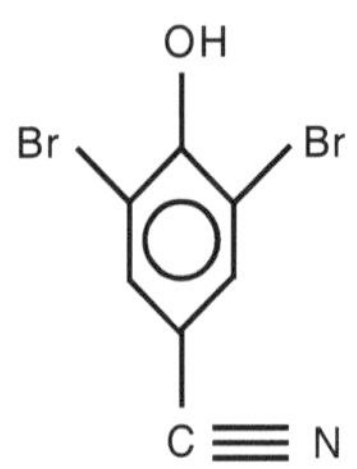

Bromoxynil

such as oats, rye, triticale, barley, wheat, maize, jowar, etc. It is visualized as potent weedicide to use at roadsides, rail road shoulders and industrial sites, etc. for young stages of weeds.

Ioxynil

Chemically it is 4-hydroxy-3,5-diidobenzonite. It was synthesized in Germany in 1896. But its weedicidal properties were reported in 1963. It is a potent weedicide for use in the crops like rice, barley and wheat as POST application. It controls a very large number of weeds belonging to the families, Fumariaceae, Labiatae, Scrophulariaceae,Caryophyllaceae and Violaceae. It is better than bromoxynil for use in agro-ecosystems.

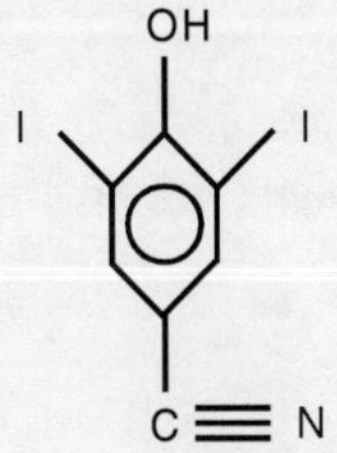

Ioxynil

Oxadiazoles

Oxadiazon

It is an useful weedicide in the crops such as onion, garlic, potato, sunflower, vegetables, soyabean, rice, groundnut, cotton, sugarcane, banana and many plantation crops like tea, rubber, etc.

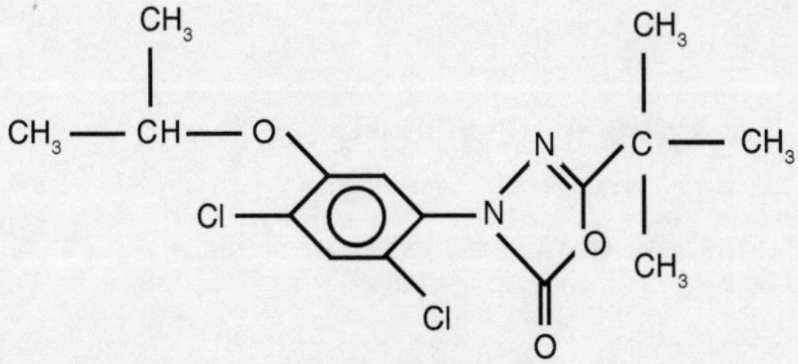

Oxadiazon

It is applied at PRE with 0.50 to 4.00 kg/ha. It has special importance to control the weeds, rhizomes and deep-rooted plant propagules. It controls *Oxalis* spp., *Amarathus* spp., *Digitaria* spp., *Uritica* spp., *Poa annua, Rhichardia scabra, Mullugo* spp., etc.

Phenols

Characters

1. Active by contact action.
2. First used in 1932.
3. Examples : DNOC, Dinoseb, etc.

Dinoseb

Chemically it is 2-sec-butyl-4,6-dintrophenol. Its herbicidal properties were first reported in 1945. Its sodium salt is most toxic and applied at POST against several kinds of weeds in crops like plantation orchard, peas, beans, potato and legumes etc. It acts as general contact weedicide.

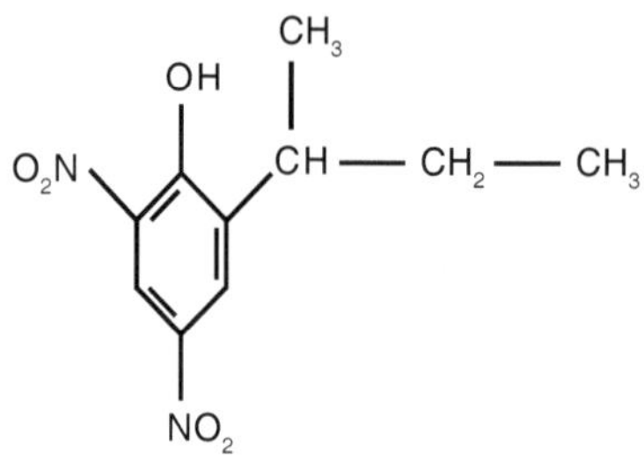

Dinoseb

Phthalamates

Characters

1. Plant shows strong epinosty.
2. Produce strong antigeotropic response.
3. Abolishes normal curvature of roots.

Naptalam

Chemically it is 2-(1-hapthalenylamino)carboxyl]benzoic acid. It was investigated by Unirogal in 1947 but it was released as weedicide in 1956. It is useful weedicide in the crops like melon, cucumber, watermelon, soyabean, groundnut, etc. It is applied PRE or POST at 2.0 to 4.0 kg/ha. It controls several broad leaved weeds including *Amarathus* spp., *Ambrosia artemisifolia, Chenopodium album, Solanum nigrum* and *Xanthium strumarium* and annual grasses like *Setaria faberi* and several others. Its field half life is 14 days. The residues of Naptalam are not deterimental to crops planted in next season. It acts as auxin inhibitor.

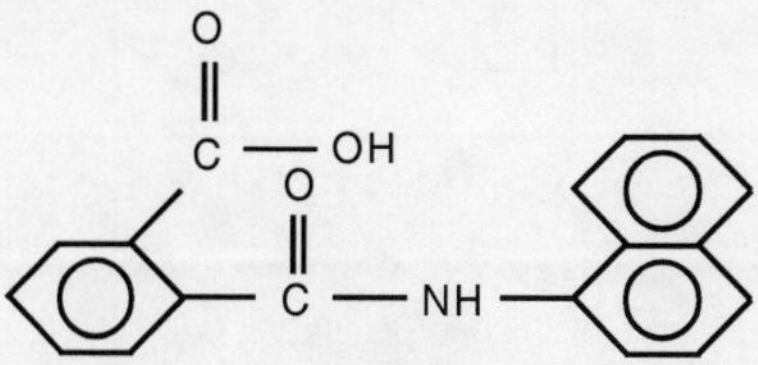

Naptalam

Pyridines

Characters

1. Mostly soil applied herbicides.
2. Acts as root growth inhibitor.
3. Cause root swelling at tips.
4. Seed germination is not affected.
5. Examples include Dithiopyr, Thiazopyr.

Dithiopyr

Chemically it is 5,5-dimethyl-2-(difluoromethyl)-4-(2-methyl propyl)-6-(trifluoromethyl)-3,5-pyridine-dicarbothioate. Mosanto first introduced this weedicide under the trade name "DIMENSION" and "STAKE OUT". But it was purchased in 1994 by Rohm and Haas. It is effective weedicide for use in the crops like rice, turf, orchard and ornamentals and several other perennial crops. It can be applied as PRE at 250-500 g/ha or at POST with 120-500 g/ha. It controls *E.c.galli, Oxalis* spp., *Digitaria* spp., *Euphorbia* spp. and *Eleusine indica* and several other annual grasses and broad leaved weeds. Its field half life is 17 days. It acts as microtubule assembly inhibitor.

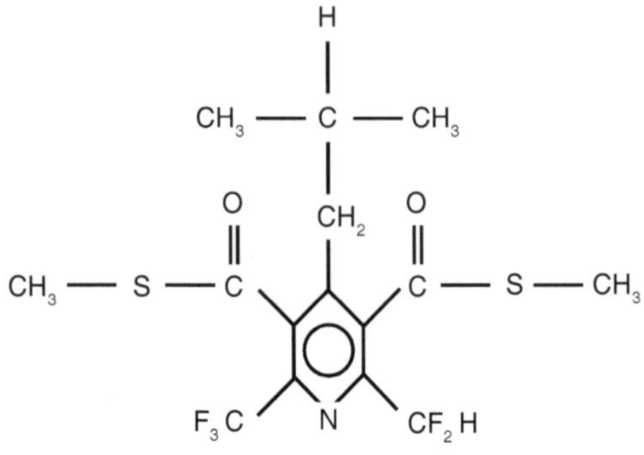

Dithiopyr

Sulphonyl Ureas

Characters

1. Investigated in 1980.
2. Most widely used herbicides.
3. Acts as growth inhibitor in weeds.
4. Causes chlorosis within 3-5 days and then necrosis.
5. Causes stunting in weeds.
6. Examples include Metsulfuron, Halosulfuron, Bensulfuron, Tribenuron, etc.

Chlorimuron

Chemically it is ethyl-2-[[[[(4-chloro-6-methoxy-2-pyrimidinyl)amino]carbonyl]amino]sulphonyl]benzoate. It was first investigated and tested in the field in 1982. But it was marketed in

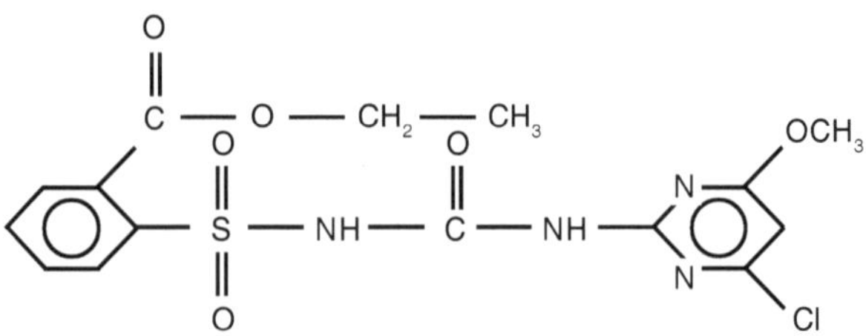

Chlorimuron

1986 under the trade name "CISSIC". It is an effective weedicide in the crops like groundnut and soyabean at POST with 8 to 13 g/ha. *Xanthium strumarium, Datura stramorinum, Amaranthus* spp., *Ambrosia* spp., *Polygonum* spp. and many other broad leaved weeds can be controlled successfully with chlorimuron. Its field half life is 40 days. Its residues last longer in high pH soil. It can be used as mixture with linuron, metribuzin, etc.

Prosulfuron

Chemically it is 1-(4-methoxy-6-methoxyl-triazin-2-yl)-3-[2-(3,3,3-trifluoropropyl)-phenylsulfonyl]-urea. Its weedicidal field activities have been tested in 1988 and registered for use in USA in 1995. It controls the annual broad leaved weeds like *Amaranthus* spp., *Ambrosia artemissifolia, Abutilon theophrasti, Xanthium strumarium, Kochia scoparia,* etc in the crops such as maize, jowar, wheat, barley, other gaminaeous crops and sugarcane when applied as POST at 15 to 40 g/ha. For increasing efficacy surfactant or oil adjuvant is necessary. Its field half life is just 10 days and its residues are moderately persistent in soil.

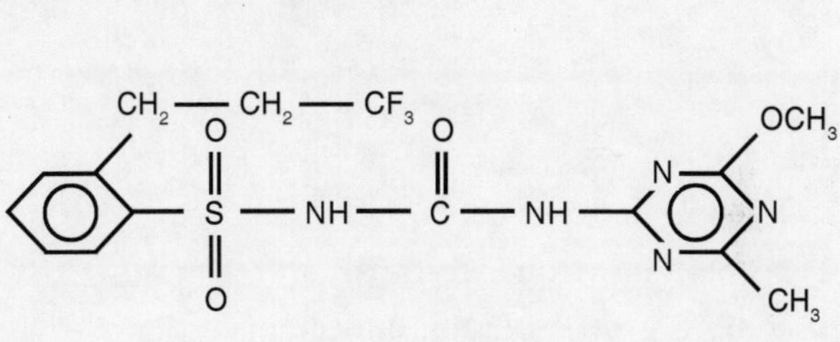

Prosulfuron

Thiocarbamates

Characters

1. Toxic to germinating weeds.
2. May injure sensitive crops like jowar, maize.
3. Shows cupped or crinkled leaves with thick, leathery texture.
4. Volatile in nature hence need soil incorporators.
5. Examples include Diallate, Butulate, Monilate, Triallate, etc.

Butylate

Chemically it is 5-ethyl bis(2-methyl propyl)carbamothiate). It was investigated in 1962. It can be applied PPI at 3.0 to 6.0 kg/ha in maize. It is effective weedicide against *Setaria* spp., *E.c.galli, Panicum dichotomiflorum, Sorghum bicolor* and *S. halepense*. It also controls certain broad leaved weeds like *Cyperus* spp within 4-7 weeks. Its field half life is 13 days. However it must be incorporated into the soil since it is volatile. It can be mixed with liquid or dry bulk fertilizers or applied through irrigation. Its synthesis is possible by reacting ethyl chlorothiolformate with abase and di-isobutyl amine.

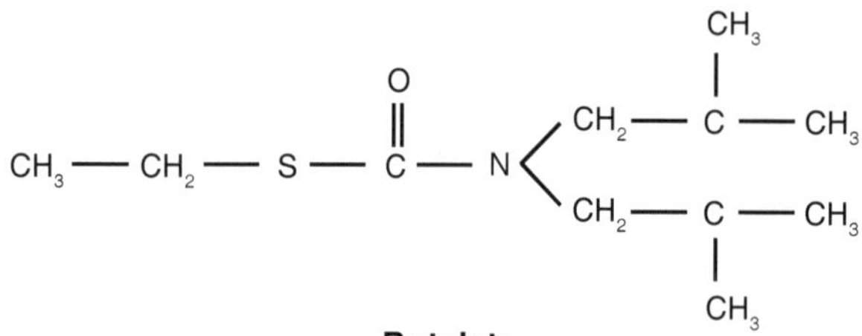

Butylate

Uracils

Characters

1. Investigated in 1962.
2. Plants shows foliar chlorosis and necrosis.
3. Inhibits growth of roots and shoots.
4. Examples includes Bromacil, Terbacil, UCC-C4243, etc.

Terbacil

Chemically it is 5-chloro-3-(1,1-dimethylethyl)-6-methyl-2,4-(1H, 3H)-pyrimidinedione. It is an effective weedicide in the crops

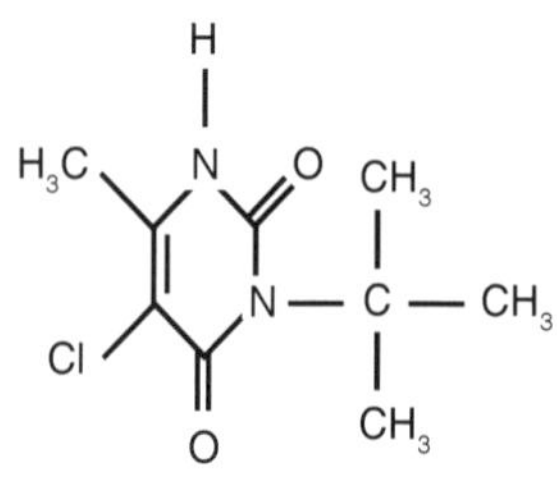

Terbacil

such as sugarcane, alfalfa, apple, peach and citrus, etc. It controls the weeds like *Lamium amplexicaule, Lolium* spp., *Descurainia* spp., *Stellaria media,* etc. when applied as POST at 1.0 to 1.5 kg/ha. Its field half life is 120 days. It is long to medium persistent weedicide in soil.

Bromacil

It controls many annual and perennial grasses and broad leaved weeds in the crops like pineapple, citrus and non-crop areas. Its field half life is 60 days. It is mixed with diuron and other compounds for the effective use.

UCC-C4243

It controls several broad leaved weeds and annual grasses. It is an effective weedicide in the crops such as lentil, barley, pea, wheat, cotton, potato, etc.

Ureas

Characters

1. Seedlings of susceptible plants become chlorotic.
2. Foliar application cause interveined chlorosis.
3. Causes yellowing of the leaf margins.
4. Older leaves are damaged more than younger ones.
5. Investigated shortly after Second World War.
6. Examples include Linuron, diuron, Tebuthiuron, Fluometuron, etc.

Fluometuron

Chemically it is N,N-dimethyl-N-[3-(trifluoromethyl) phenyl]urea. It is investigated first in 1964. In cotton, jowar, sugarcane, ornamental plants and citrus, it can be applied PPI, PRE at 1.00 to 1.5 kg/ha and as E-POST at 1.00 to 2.00 kg/ha for controlling several broad leaved weeds and grasses like *Digitaria* spp., *Setaria*

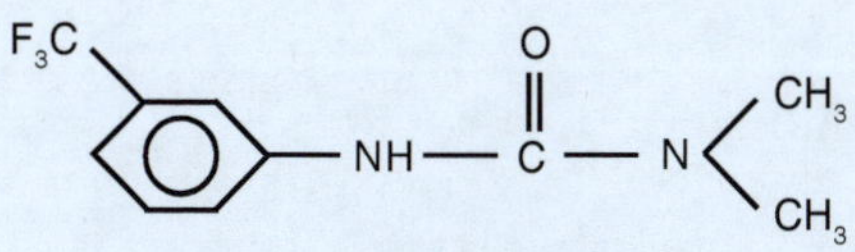

Fluometuron

spp., *Ipomea* spp., *Portulaca* spp., *Polygonum* spp., *E.c.galli, P. dichotomiflorum, Brachiaria platyphylla, Cessia obtusifolia, Euphorbia esula, Chenopodium album, Eleusine indica,* etc. Its field half life is 85 days. It can be applied in mixture with MSMA. Fluometuron can be synthesized by alkylamination of the appropriate aromatic isocynate. It acts as photosynthesis inhibitor in weeds.

Diuron

Chemically it is N-(3,4-dichlorophenyl)-N,N-dimethyl urea. It was first discovered in 1951. It is an effective herbicide for use in the crops such as cotton, grapes, coffee, tea, pineapple, apple, pear, many plantation crops and orchard crops. At 0.8-2.5 kg/ha when applied as PRE, it controls several broad leaved weeds. At POST application with 0.2 to 0.4 kg/ha, it controls the weeds in jowar and maize. Its field half life is 60 days. By reacting 3,4-dichlorophenyl isocyanate and dimethylamine diuron can be synthesized. It acts as photosynthesis inhibitor.

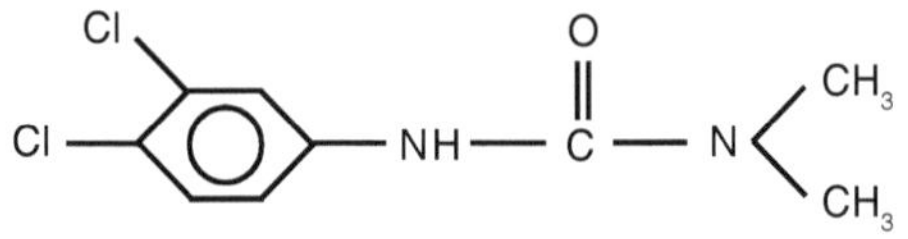

Diuron

Bioherbicides

There are three bioherbicides currently released under the trade name "BIOMAL", "COLLEGO" and "DEVINE". BIOMAL is effective against the weed *Malva pusilla*. Biomal is available as wettable powder. COLLEGO is effective against the weed, *Aeschynomene virginica* in rice and soyabean crops. DEVINE is effective against *Morrenia odorata* in citrus crop.

Chapter 12

PLANT NUTRIENTS AND DEFICIENCY

A knowledge on plant nutritional requirement, quality of soil and characteristics of fertilizer is essential for hoping maximum yield of the crops. As essential plant nutrients, at least sixteen chemical elements are necessary for the growth and development of all plants. Carbon, hydrogen, oxygen are received to the plant through air and water. Other nutrients are collected by the plants through soil, fertilizer and manures.

Essential nutrients are needed for normal growth and their deficiency can not be prevented or corrected by adding other nutrients or components. The list of essential mineral nutrients required for crops is given in Table 16.

Table 16: List of Essential Mineral Nutrients for Crops

Sr.No.	*Element*	*Symbol*	*Forms Absorbed*
1.	Nitrogen	N	NH_4^+, NO_3^-
2.	Oxygen	O	—
3.	Phosphorus	P	$H_2PO_4^-$, $HPO_4^=$
4.	Potassium	K	K^+
5.	Sulphur	S	$SO_4^=$
6.	Calcium	Ca	Ca^{++}
7.	Magnesium	Mg	Mg^{++}
8.	Carbon	C	—

contd...

Table 16–Contd...

Sr.No.	Element	Symbol	Forms Absorbed
Trace Elements (Micronutrients)			
9.	Boron	B	H_3BO_3, $H_2BO_3^-$
10.	Iron	Fe	Fe^{++}
11.	Manganese	Mn	Mn^{++}
12.	Copper	Cu	Cu^+, Cu^{++}
13.	Zinc	Zn	Zn^{++}
14.	Molybdenum	Mo	$MoO_4^=$
15.	Chlorine	Cl	Cl^-

In general, plant nutrients are categorized into two groups:

1. Primary, Secondary and Micro or
2. Major and Minor elements or Macro and Micronutrients

The elements those required in large amount are called major or macro and those required in small amount are called micronutrients. The primary elements are N, P and K and the secondary S, Ca and Mg. However, the above six elements are macro elements. Though all these nutrients are essential to all crops, their requirement differs from crop to crop or plant to plant.

Plant Nutrients and Deficiency

Each essential nutrient has its individual tasks to do inside the plant body. N needs of a crop can not be met by P application or Mg. In deficient plants the concentration of a nutrient falls below a certain critical level resulting in disturbance in its vital functions. Thus, the nutrient deficient plant appears abnormal or shows some special features. Due to nutritional problems, deficiency symptoms are seen in plants. The role of nutrients and the deficiency characteristics in crops are given below.

Nitrogen (N)

Nitrogen is required for vigorous growth, branching or filtering, leaf production and size enlargement of a plant and also responsible for dark green colour of leaves and stems. Nitrogen is available as constituent of NO_3^-, NH_4^+ and also essential nucleic acids, coenzymes and porphyrines such as chlorophyll and cytochromes. The deficient plants get stunted and becomes yellowish green. Due to synthesis of anthrocyanin, purple colour appears to leaves. In severe case leaves turn brown and die, vegetative growth is stunted, flowering and fruiting bodies are affected thus, crop yield is adversely affected.

Potassium (K)

It is essential for photosynthesis, respiration and chlorophyll synthesis, etc and available in fixed and exchangeable forms (K^+). Potassium is an activator for at least 40 enzymes and maintains electroneurality in the cells. 'K' deficient plants shows following features:

1. Leaves become reddish dark brown.
2. Marginal chlorosis.
3. Necrosis to old leaves begin.
4. Leaf tips become curly.
5. Leaf margins turn upward.
6. Stunting of growth.
7. Inter nodes become very short.

Sulphur (S)

Sulphur is available as SO_4^{--} in the form of organic matter through microbes and also from minerals. It is a part of cysteine, cystine and methionine thus, essential for protein production. In cruciferous plants it is quite abundant. It is involved in chlorophyll formation and enzyme activations. It is also a part of vitamins like biotin and thiamine (B_1) and coenzyme A. Sulphur deficient crop shows following features:

1. Leaves become pale yellow or light green.
2. Produce anthocyanin.
3. Chlorosis appears on young leaves.
4. Growth of crop retarded.
5. Nodulation becomes poor in legumes.
6. N-fixation reduced.
7. Stems become thin and slender.
8. Maturity delayed in cereal crops.
9. In oil seed crops, oil production is significantly reduced.

Phosphorus (P)

It is available in organic and inorganic form and essential for growth, cell division, lengthening of roots and development of seeds and fruits. Phosphorus is also useful in ripening of the fruits. It is essential as part of nucleic acids, ADP, ATPs, NADP, NAD and phoshpolipids. Deficient crop shows following characteristics:

1. Leaves become purple and red.
2. Fruits ripen slowly.
3. Crop becomes dwarf.
4. The walls of xylem and phloem become thin.
5. Stunting in young crops.
6. Leaves fall down prematurely.
7. In older leaves symptoms start first.

Calcium (Ca)

It can be made available through exchangeable form, cations absorbed on clay micelles. Major portion of calcium is in non-available form. It is required as an activator for some enzymes. It acts as buffer to neutralize acids. It has important role in cell division, growth and root lengthening. It also acts as inhibitor of enzymes. Following symptoms are noted in calcium deficiency of crops:

1. Leaves become small, distorted, cup shaped, crinkled/ cholorotic.
2. Leaves colour becomes dark green.
3. Chlorosis seen with marginal cells.
4. Fruiting significantly affected (No fruiting or very less)
5. Symptoms observed first on growing tips and the youngest leaves.
6. The disease is called "dieback" since young meristematic breakdown occurs first and then older ones.

Magnesium (Mg)

It is very important for photosynthesis of plants. It occupies the centre spot in the chlorophyll molecule. It has important role in activation of enzymes, energy transfer, production of proteins, metabolism of carbohydrates, maintenance of electrical balance, etc. The deficient plant shows following characters:

1. Interveinal chlorosis in older leaves.
2. Leaves become small and brittle.
3. Leaves drop down early.
4. Twigs become weak.
5. The veins remain green but leaf area turns yellow.

Boron (B)

It is exchangeable and in soluble form, found more in organic soil. It is available as Bo^{++}. It is essential for carbohydrate transport. It is also essential for germination of pollen grains. Boron has a very important role in membrane stability and cell wall development, cell division and cell extension.

Boron deficiency symptoms

1. Affect the growing point of shoot, root and young leaves.
2. Young leaves deformed in cluster.
3. Leaves become coppery and thick.
4. Cracking of cork formation in stems, fruits and stalks.
5. Reduction in bud, flower and seed production.
6. Formation of hollow heart in ground.
7. Cork formation in apple, heart rot in sugarcane, etc.

Copper (Cu)

It is essential in chlorophyll formation and an important constituent of several enzymes like cytochrome oxidase, ascorbic acid oxidase and phenolase. It is also part of plastocyanin and plastoquinone. Copper is involved in lignin formation and protein and carbohydrate metabolism.

Deficiency symptoms

1. Entire plant shows witted appearance.
2. Exanthema disease caused in fruit trees.
3. Leaf necrosis on young leaves, found along the margins.
4. Shoot tips become narrow.
5. Leaves become twisted.
6. Affects yield very significantly.

Chlorine (Cl)

Chlorine plays important role in production of oxygen during photosynthesis. For raising cell osmotic pressure and maintaining tissue hydration, chlorie is considered to be involved in the process.

Deficiency symptoms

1. Chlorosis in younger leaves.
2. Wilting of plant due to transpiration.

Iron (Fe)

It is present in sufficient quantity however, may not be in the available form. It is abundant in acidic soil rather than in neutral or alkaline. It is available in ionic form (Fe^{++} & Fe^{+++}) and essential for making of chlorophyll, carbohydrate production, respiration, reduction of nitrate and sulphate in N assimilation. It is also essential for flavoprotein and cytochrome synthesis. Iron deficiency symptoms recorded in certain crops are as follows:

1. Symptoms first seen on younger leaves.
2. Yellowing of interveinal area leaves.
3. Leaves become pale white.

Manganese (Mn)

It is found dissolved in soil solution. It is also available as an exchangeable ion absorbed on the soil particles. Manganese is essential for splitting the water molecule during photosynthesis. It has an important role in N-metabolism and carbon-dioxide usage for photosynthesis. It is responsible for activation of many enzymes like malic dehydrogenase etc.

Deficiency symptoms

1. Leaf veins become green.
2. Interveinal area shows mottled chlorosis.
3. Shows grey lesions (off colour wounded tissue) on the leaves.
4. Stems become yellow and many times hard and woody.
5. Symptoms appear first on young leaves.

Molebdenum (Mo)

It is found in dissolved form and also found as a form adsorbed on the soil particles. It is available as Mo^{++}. It can also be located in a non-available form. Molebdenum is essential for N-fixation in root nodules of legumes and nitrate assimilation. It directly participates in protein synthesis. It is a good ingredient for ascorbic acid synthesis.

Deficiency symptoms

1. In folding of leaves or cupping of leaves.
2. Features resembles with N-deficiency.
3. Cause marginal scorching/necrosis.
4. Chlorotic interveinal mottling of lower leaves.

Zinc (Zn)

It is found in the soil at low concentration. However, it is more readily available in acidic soils. Zinc is required for several enzyme systems, auxins, protein synthesis and seed production, etc. It is believed that zinc promote nucleic acids needed for protein synthesis.

Deficiency symptoms

1. Stunted growth.
2. Smaller leaves (little leaf disease).
3. Chlorotic bands on either side of the midrib in monocots.
4. Brown rusty spots on leaves (rice, etc.)

Soil

The efficacy of fertilizer is dependent on the quality of soil. Thus, soil decides to which fertilizer nutrients are converted, conserved and conveyed to the roots. About 45-50% of soil's volume is made up of mineral matter, 1-5% is organic matter and the remaining 50% consists of open pore spaces of air and water (Fig. 27).

All soil nutrients are present in soils. Nutrients are inherited either from various rocks and minerals or from organic matter and past additions. Most of the nutrients are available as a part of organic and inorganic (minerals) matter. The organic and inorganic nutrients co-exist and can be converted into one another in the soil.

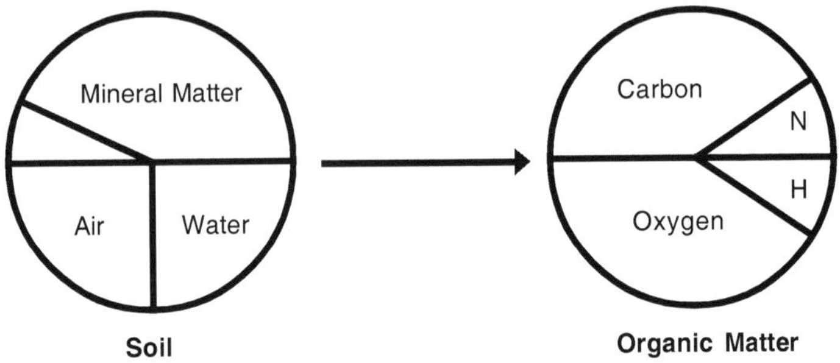

Fig. 27: Composition of Soil.

The mineralization is carried out by certain micro organisms (bacteria, etc.). The rate of mineralization depends on the type of organic matter and environmental conditions. In flooding the rate is slowed down, but, in near-neutral soil it is fastest.

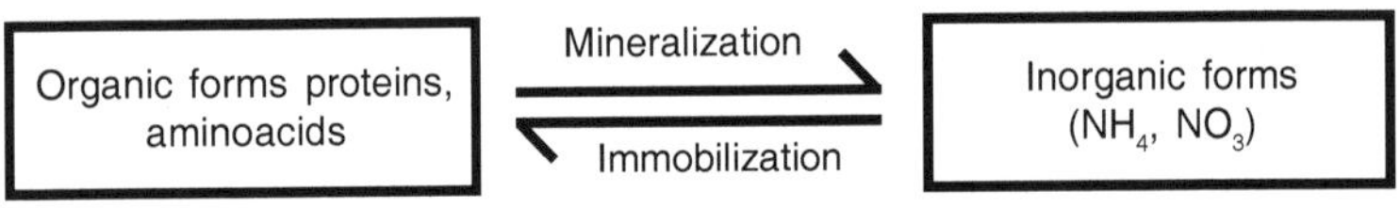

Chapter 13

CHEMICAL FERTILIZERS

Chemical fertilizers have occupied an important place in modern agriculture. The use of chemical fertilizers is increasing day by day. During 1992-93 Indian farmers purchased fertilizers worth of Rs.9952 crores and of approximately 1000 million fertilizer bags of 50 kg. In India, the first single super phosphate (SSP) factory was set up in 1906 at Ranipet (Madras). Fertilizers are site specific and area specific agrochemicals.

Definition

"It is a mined or manufactured material containing one or more essential plant nutrients in immediately or potentially available forms in commercially valuable amounts for safety use for promoting the growth/development of a plant". Out of the 15 individuals mentioned under macro and micro nutrients, a fertilizer must constitute one or more members. A true fertilizer is a suitable form of minerals. It may be products made from raw materials in a factory like Urea, DAP, SSP, complexes, etc. or process like MOP / or rock phosphate. Urea contain N, SSP contains P, S & Ca; PAP contains N and P. Likely, complex 12-32-16 and 17-7-17 contains NPK and zinc sulphate contain Zn and S. P in the rock phosphate is insoluble at first stage but, in acid soils it is changed to the soluble form which is used by the plant. In addition to above characters, a fertilizer should have no harmful or poisoning or toxic effect on either plants or animals or humans. In soil all nutritional elements are present.

However, fertilizers are added only to increase the supply of essential plant nutrients to fulfil the demand of the crop. The fertilizers are only carriers of plant nutrients. Hence, they should be used according to need. The selection of fertilizer is dependent on environment, quality of soil and plant species.

Air is an important source of N in all fertilizers which contains 79% N. Rock phosphate and sulphur is used in phosphate fertilizers. Thus, most raw materials used for fertilizers are also natural ones. The organic and inorganic sources are found changing within the same nutrient cycle. However, they bound together by some processes like mineralisation and immobilization. The fertilizers should be used with great care that

1. They may be used as per the need only.
2. They should be used efficiently.
3. Avoid excess application and
4. Treat organic vs. inorganic.

The following Table 17 shows nutrient content of fertilizers.

Table 17: Nutrient Content in Fertilizers and Manures

Sr.No.	*Fertilizer*	*Nutrient Content (%)*				
		N	*P_2O_5*	*K_2O*	*S*	*Others*
1.	Ammonium chloride	20	-	-	-	-
2.	Ammonium sulphate	21	-	-	24	-
3.	Amm. Phos. Sulphate	16	20	-	15	-
4.	Amm. Phos. Sulphate	20	20	-	15	-
5.	Bone meal stemed	1.5	25	-	-	-
6.	Borax	-	-	-	-	B 10.5%
7.	Calcium Amm. Nitrate	25	-	-	-	Ca
8.	Caster cake	5.6	1.8	1.0	-	-
9.	Copper sulphate	-	-	-	13	Cu 24%
10.	Chelated Zinc as Zn-EDTA	-	-	-	-	Zn 12%
11.	Chelated Iron as Fe-EDTA	-	-	-	-	Fe 12%

Contd...

Table 17–Contd...

Sr.No.	Fertilizer	Nutrient Content (%)				
		N	P_2O_5	K_2O	S	Others
12.	Diammonium phosphate	18	46	-	-	-
13.	Dried blood	10	1.2	0.7	-	-
14.	Groundnut cake	7.0	1.5	1.3	-	-
15.	Gypsum (Aqric-grade)	-	-	-	13	Ca 16-19%
16.	Magnesium sulphate	-	-	-	13	Mgo 16%
17.	Manganese sulphate	-	-	-	15	Mn 30.5%
18.	Nitrophosphate	20	20	-	-	-
19.	Nitrophosphate	23	23	-	-	-
20.	Nitrophosphate	15	15	15	-	-
21.	NPK Complex	12	32	16	-	-
22.	NPK Complex	10	26	26	-	-
23.	NPK Complex	17	17	17	-	-
24.	NPK Complex	19	19	19	-	-
25.	NPK Complex	14	35	14	-	-
26.	Neem cake	5.2	1.0	1.4	-	-
27.	Phosphygypsum	-	-	-	16	Ca 21%
28.	Potassium chloride	-	-	60	-	Cl
29.	Potassium sulphate	-	-	50	18	-
30.	Rapeseed cake	5.1	1.8	1.1	-	-
31.	Rural compost	0.6	0.5	-	-	Ca
32.	Rock phosphate	-	18.20	-	-	Ca
33.	Single superphosphate	-	16	-	12	Ca 20%
34.	Solubor	-	-	-	-	B 19.00%
35.	Sludge (activated)	58	3.2	0.6	-	-
36.	Urea	46	-	-	-	-
37.	Urea Amm. Phosphate	28	28	-	-	-
38.	Urban compost	1.5	1.0	1.5	-	-

Classification

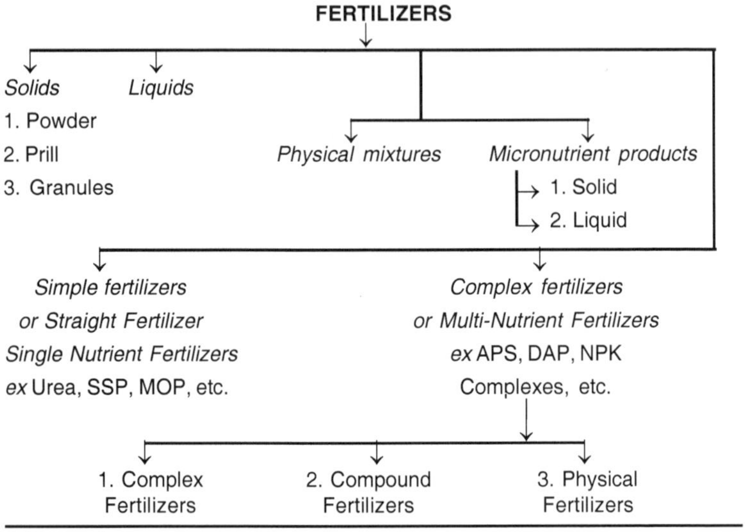

India is agricultural country. Hence, there is tremendous potential for fertilizer production and marketing. More than 25 fertilizers are produced in India. On the global scenario, India ranks 3rd in fertilizer consumption. Uttar Pradesh is using largest amount of fertilizer in India. However, Punjab using most fertilizers per hectare. While, Kerala is engaged in highest usage of potash as N. The contribution of fertilizers to increase in food production is 50%. Hence, it is needless to say that the use of fertilizers is mandatory part of modern agriculture. Balanced use of fertilizers correct all deficiencies and promote the growth and yield of the crops.

Simple or Straight Fertilizers

Simple fertilizers are those which contain only one nutrient out of N, P and K. Example: SSP, Urea, etc.

Multi Nutrient Fertilizers

Multinutrient fertilizers are those which contain N+P or N+K or P+K or N+P+K. Example: NPK complexes (PPK 12-32-16), APS, DAP, etc.

Compound Fertilizers

The term compound fertilizer is used for complexes and granulated mixtures.

Physical Fertilizers/Mixtures

Physical fertilizer is mixing of fertilizers to get a mixture of appropriate grade.

In a complex fertilizer, different nutrients are reacted in a factory before the final product stage. Example: NPK complex (12x32x16). In this fertilizer, DAP is combined with MOP as DAP is source of N & P and MOP is source of K. Physical mixtures (powders) do not have the uniformity in composition within the bag or a lot since their ingredients are not bound together and can separate out due to differences in size, weight and density. This may be avoided by granulating the mixture. On the fertilizer bags, accounting the number like 18-18-18 or 15-15-15 refers to N-P_2O_5-K_2O respectively. The middle figure refers to P_2O_5 right hand refers to K_2O and left to N.

Types of Fertilizers

Urea

Urea is most common source of N which contain 46% N and no other mineral nutrients. It is readily soluble in water and widely used in agroecosystem for increasing the yield of crops. Nitrogen available in the amide form (NH2) is changed to ammonium (NH4) which is absorbed easily by the plant roots. The leaves of certain plants can absorb amide directly through foliar spray. However, amide form nitrogen is absorbed by the roots at slow rate. It is most effective favourable fertilizer and widely used for cereals, pulses and many cash crops like sugarcane, tobacco and cotton etc at a appropriate situation.

The urea in soil is converted into ammonia by the enzyme urease which is present in soil. The fertilizer reaction of urea is represented below.

Ammonium Chloride (ACl)

As ammonium chloride does not contain a basic cation, it is acid forming fertiliser and readily soluble in water. It contains 25%

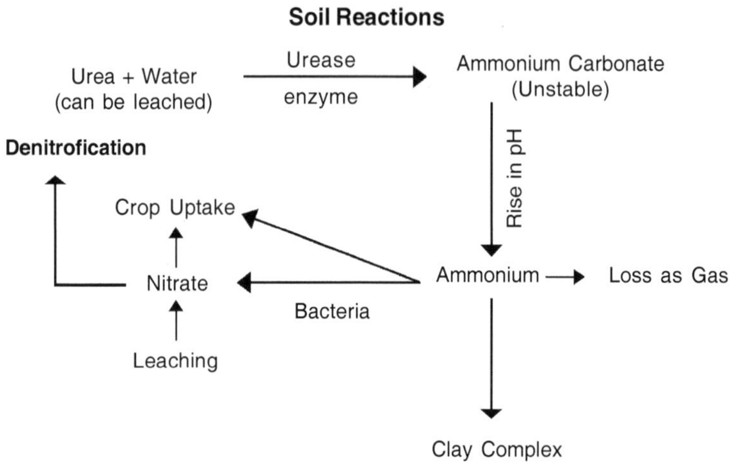

nitrogen and rest is Cl. However, its N is available in the ammoniacal form in plants. It is effective fertilizer for many agricultural crops.

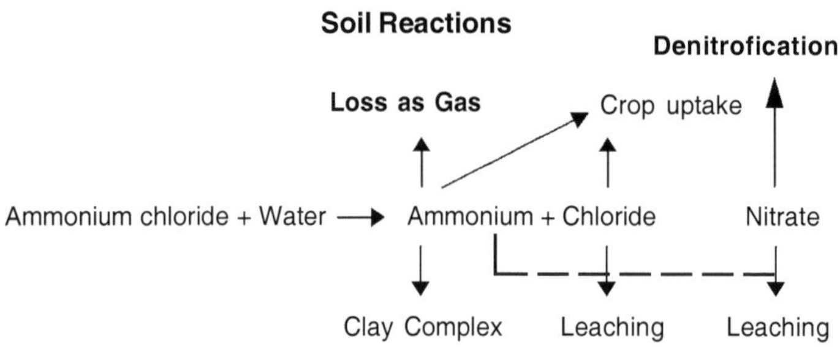

Ammonium Sulphate (AS)

Ammonium sulphate is white crystalline material. Its moisture absorbing capacity is less than urea. Hence, it is easy to store. Its colour may be changed due to production process. However, its nutritive value remains same. Ammonia contains 21% N and 24% S. In recent years, N% in fertilizer is very less, only 2%. It is readily soluble in water. Its continuous use lowers soil pH since it is an acid forming fertilizer. However, by the lime application pH can be cleared. Ammonium chloride and ammonium sulphate produce about 3 times more acidity than urea. The soil reactions are represented below:

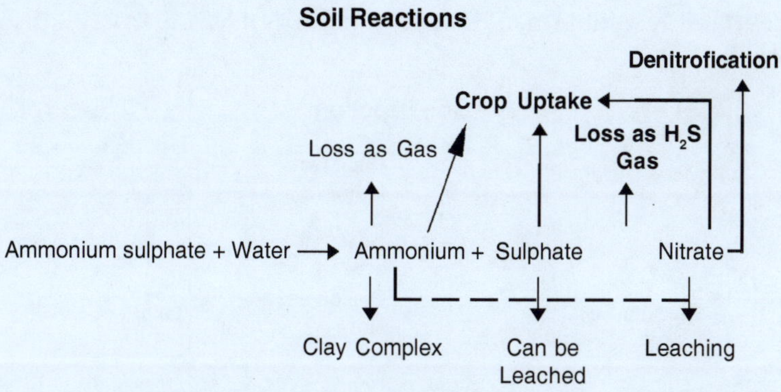

Calcium Ammonium Nitrate (CAN)

It contains calcium ammonium and nitrate. CAN is in the form of grey round prills and granules. In CAN, nitrogen comprises 25-26%. Calcium comes from calcium carbonate (lime) used for mixing with ammonium nitrate for the production of CAN. CAN easily absorbs the moisture from the air but dry forms are free flowing ones. The soil has no acidic effect due to the application of CAN. Its nitrogen part is readily soluble in water but, this does not happened with calcium.

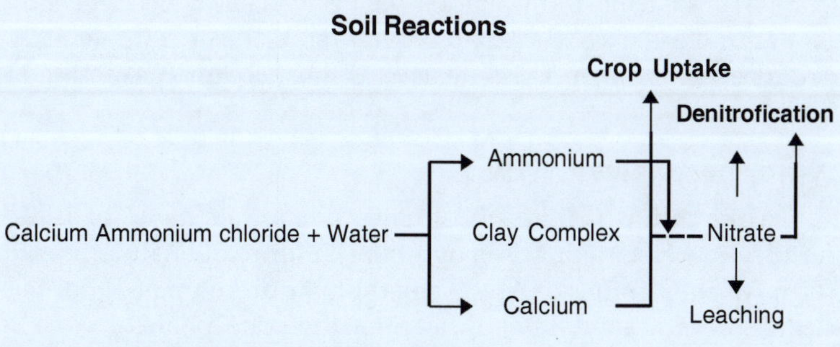

Single Super-phosphate (SSP)

The grayish coloured SSP is a very popular fertilizer which contains 16% P_2O_5, 12% S and 20% Ca. It is in use for last more than

150 years. P and K of SSP are derived from rock phosphate while, S is from sulphuric acid used during its manufacture. SSP is available in the market as a powder or granules. Its S and Ca are comparatively less readily soluble than P in gypsum. SSP is most important carrier of S.

Soil Reactions

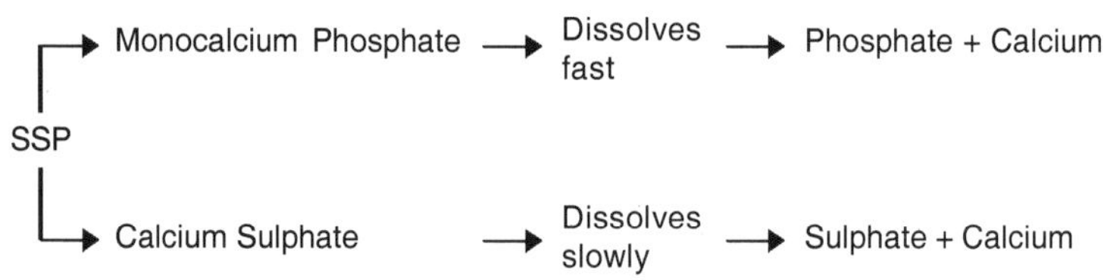

Di-ammonium Phosphate (DAP)

It is grey coloured granular fertilizer which contains 18% N + 46% P_2O_5. It is common source of P and next to urea in India with respect to N source. It is readily soluble in water and having both its nutrients in plant available form. It can be used in combination of other fertilizers. However, it is very much suited for basal application where higher rate of P application is recommended.

Urea Ammonium Phosphate (UAP)

It contains 68% N in the amide (urea) form and 32% in ammonium form. Entire nutrients of this fertilizer are readily soluble in water in available form, amide-N being available after conversion into ammonium. UAP-28-28-0 is the first fertilizer commercially produced in the world and hence it is most common example of UAP.

Nitrophosphates (ANPS)

Ammonium nitrate phosphates is another name of these fertilizers. ANPS' half N is in ammonical form and half in nitrate form. Various grades of ANPS have similar make up of N but the degree of water solubility is different in their phosphates.

Examples

ANP - 15-15-15 has 30% water soluble phosphate,

ANP - 20-20-0 has 60% water soluble phosphate, and

ANP - 23-23-0 has 80-85% water soluble phosphate.

Rock Phosphate (RP)

Mussoorie rock phosphate and Purulia rock phosphate are the common forms of rock phosphates found in India. P_2O_5 content of Purulia rock phosphate comes about 18%. Rock phosphate is used as a raw material in manufacturing fertilizers, but some grades are used directly as fertilizers in acid soils. Rock phosphates have little water soluble P. However, in acid soils, their insoluble P is solubilized by the action of soil acids and such form is useful for plants. Rock phosphate also contain calcium and have some limiting effects.

Soil Reactions for Phosphatic Fertilizers

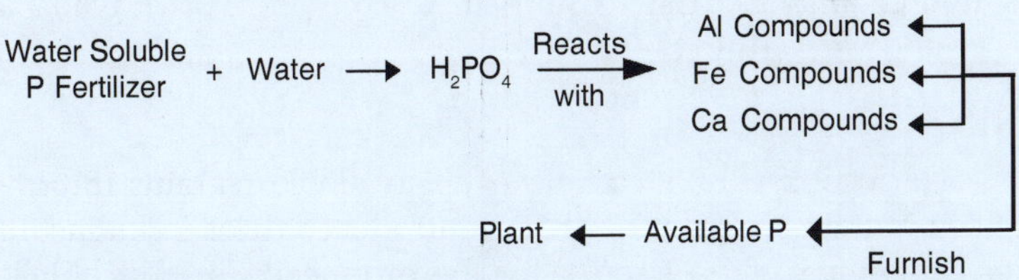

Dinitrophosphates

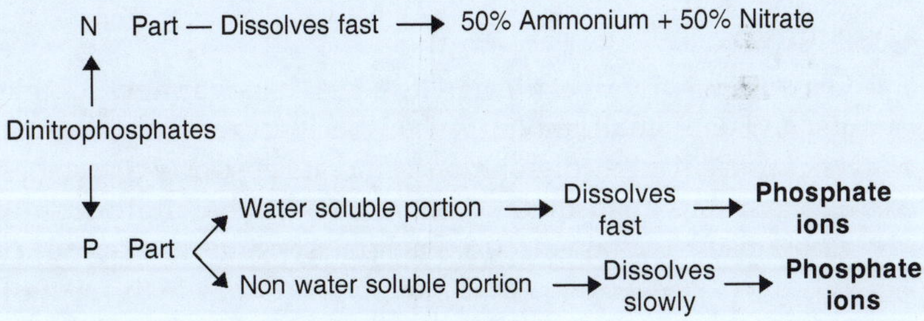

Potassium Chloride (KCl or MOP)

KCl have elemental composition of 50% K and 50% Cl. However, its K_2O content is 60%. Sometimes, its white colour is changed to red. MOP is readily soluble in water and its K is in plant available form.

Potassium Sulphate (SOP)

It is marketed either as a crystalline powder with a very little dust or in granular form. It contains 50% K_2O and 18% S. Both nutrients are readily soluble and in plant available form. In damp condition, potassium sulphate stores well due to its excellent physical properties. SOP is manufactured by the action of sulphuric acid with potassium chloride (MOP) or by reacting other compounds. Fertilizer SOP is virtually chloride free, does not add to the salinity hazard and has a low salt index.

Ammonium Phosphate Sulphate (APS)

Ammonium phosphate sulphate is a complex of ammonium sulphate and ammonium phosphate. There are two grades, APS - 16-20-0 and APS - 20-20-0, both contain 15% S which comes from ammonium sulphate portion. Out of these two grades APS - 20-20-0 is more common. All the tree nutrients are present in plant available form and are readily soluble.

NPK Complexes

The fertilizers NPK complexes are available in granular forms which contain N, P and K. MOP is introduced during production for combining with N and P for the production of NPK. These nutrients are water soluble and in plant available form with some exceptions (P in ANP). NPK complexes contain 45-63% plant nutrients.

Gypsum

Gypsum is sulphur campaign fertilizer. It is sparingly soluble in water and only small amount is used as a fertilizer, therefore, soil water is sufficient for its dissolution. Mineral gypsum which comes mostly from Rajasthan mines contain 13% S. The bi-product or phosphogypsum contains 16% S. Phosphogypsum is a bi-product of wet processed phosphoric acid. About 4.5 tonnes of phosphogypsum is discharged every day for a tonne of P_2O_5 in the process of production of phosphoric acid and for the end product DAP. Calcium sulphate is active compound in above types of gypsum.

Pyrites

Pyrite is a mineral found in Amjhore area of Bihar state. The S in pyrites is in sulphide form which is changed into plant available sulphate form in soil. As a sulphur fertilizer, iron pyrites are used in some parts of our country. Iron pyrites contain 18-22% S.

Magnesium Sulphate

In India, it is recognized as epsum salt ($MgSO_4.7H_2O$) and is most commonly used against Mg deficiency in crops. It contains 16% MgO and 13% S. Both are readily available forms. More concentrated forms of magnesium sulphate are in use in many countries. Example: Kieserite ($MgSO_4.H_2O$). It contains 27% MgO and 22% S.

Elemental Sulphur

Elemental sulphur is not directly absorbed by plants. It is to be changed into the plant available form in soil. Elemental sulphur consists 85-100% S. The sulphur obtained as a mineral or collected during purification of crude oil, is used as raw material for sulphuric acid production and later, manufacturing phosphatic fertilizers, ammonium sulphate or other minerals.

Zinc Sulphate

There are two grades of zinc sulphate. The heptahydrate (hepta-7) form contains 21% Zn, while, monohydrate form contains 33% Zn. The heptahydrate form is commonly used in India as it shows relatively poor physical properties and thus maintain quality.

Other Slats

Borax (B)	is sodium borate ($Na_2B_4O_7.10H_2O$) contains 10.5% B.
Solubor	is a mixture of $Na_2B_4O_7.5H_2O+$ $Na_2B_{10}O_{16}.10H_2O$.
Boronated SSP	Boron added SSP borax.
Copper Sulphate	It contains 24% Cu and 13% S.
Ferrous Sulphate	It contains 19% Fe and 15% S.
Manganese Sulphate	It contains 30.5% Mn and 15% S.
Ammonium Molybdate	It contains 52.0% Mo and some N.

Chelated Fertilizers

The chelated products helps to keep the nutrient ion in a soluble but stable form till it get absorbed by the roots. In chelated fertilizers the nutrient cations are bound in a ring like structures by organic molecules. The most common products are given below:

1. Zn - EDTA chelate
2. Fe - EDTA chelate. EDTA (Ethylene Diamine Tetra Acitic Acid) is the organic molecule working as chelating agent for binding Zn and Fe.

Table 18: Matching of Phosphatic Fertilizers with Soil and Crops

Sr. No.	*Soil Reaction*	*Sulphur Deficiency Problem*	*Crop Duration*		
			Short	*Medium*	*Long*
1.	Moderate to strong acid	Yes	SSP or SSP–Rock P mix. or SSP–FYM mix.	SSP-Rock P mix. Rock P – FYM mix. & AS for N+S or APS	Rock P+S Source
		No	Any source or mixture having 30-50% WSP	Rock P+FYM or 30% WSP nitrophos	Rock P
2.	Near Neutral	Yes	SSP or APS or any high WSP with AS or gypsum for S	SSP, ASP or source with 50-70% WSP with gypsum or AS for S	Rock P–SSP mix. or source having 30% WSP+AS or gypsum
		No	Any high water soluble P–source	Any source with 50-60% WSP	Rock P+WSP mix. or Rock P+FYM or source with 30% WSP
3.	Alkaline	Yes	SSP or APS or any water soluble P source with AS or gypsum for S	SSP, APS or 60-70% WSP source with AS or gypsum for S	SSP, APS or source with 30% WSP+AS or gypsum for S
		No	Any source with over 80% WSP	Any source with 60-70% WSP	Any source with 30-50% WSP

WSP = Water Soluble P.

The use of fertilizer in various crops increases the yield and thus uplift the socio economic status of the farmers. At least 40-50% increase in agricultural production is expected. However, in some cases it is as high as 75%. This increase is credited to fertilizers only. A significant increase in the yield of the crops like rice (Kharif and rabi), wheat (R), sorghum (K), maize (K), pearl millet (K), finger

Table 19: General Recommendations of Fertilizers

Sl. No.	*Kg/ha added*			*Yield obtained*
	N	*P_2O_5*	*K_2O*	
1.	0	0	0	15000
2.	60	0	0	2400
3.	60	30	0	3000
4.	60	30	30	3300
5.	120	0	0	2700
6.	120	60	0	3600
7.	120	60	60	4000
8.	150	90	90	3900

millet (K), groundnut (R), and cotton have been noticed due to application of fertilizers (Kulkarni, 1984). Increase in the yield also noted with the crops like chickpea, pigeon pea, groundnut and mustard (Pillai, 1985) by applying N, P_2O_5 and K_2O with different combinations. In wheat, rice, groundnut, mustard, soyabean, pigeon pea, sunflower and linseed, increase in the yield have been noticed due to the application of sulphur. Likely, wheat, rice, maize and potato showed significant increase in the yield due to application of zinc and wheat, rice, sorghum and groundnut due to application of iron.

Chapter 14

BIO-FERTILIZERS AND ORGANIC COMPOSTING

For increasing the yield and keeping the plants healthy, chemical fertilizers are used at increasing rate. Hence, there is need to adopt a strategy of integrated nutrient supply. The judicious combination of chemical fertilizers, organic manures and bio-fertilizers would worth avoiding the risk of non-renewable forms of energy. *Azotobacter, Azospirillum, Azolla* etc. are important micro-organisms and are actively engaged in nitrogen fixing and supply of nitrogen nutrition to the plants. Hence, the present chapter deals with various bio-fertilizers, organic manures and organic matter and composting.

Phosphate-Solubilizing Microorganisms

Next to nitrogen and phosphorus, microorganisms are important nutrients for plants. The soil contains high organic matter including rich organic forms of phosphorus which comes mainly by way of decaying vegetation.

Rock phosphate is one of the basic materials for phosphatic fertilizer production. In India, 100 million tonnes of rock phosphate deposits are available. Superphosphate is a common form of phosphatic fertilizer used in India.

The bacterial species belonging to the genera *Bacillus* and *Pseudomonas* possess the ability to bring insoluble phosphates in soil into soluble forms by secreting organic acids like formic, acetic, propionic, lactic, glucolic, fumaric and succinic acids. Bacterial cell size ranges from 1.1 to 2.2 µm. Rod shaped *Bacillus* is grame +ve while, *Pseudomonas* is gram –ve bacteria. Transport zones of clearing around microbial colonies indicate extent of P solubilization, spreading type growth and spore formation for *Bacillus*. Glucose is main carbon source but, can utilize other carbon source and respire aerobically.

Several fungi species belonging to the genera *Penicillium* and *Aspergillus* are also involved in solubilization of phosphates in soil. The above acids decrease the pH of soil and dissolve bound forms of phosphate. According to Oswal & Bhide (1972) some hydroxy acids chelate with calcium and iron and effectively solubilize and utilize the phosphates. Aspergillus conidial heads are coloured conidiospores pitted, appearing rough, arise separately from the substratum, 400-700 µ or 1000 µ long x 5-15 µ in diameter, most molds strictly aerobic, C-source–glucose or other sugars. The medium for fungi and bacteria is Pikovskaiya's medium.

The list of microorganisms and sources of phosphate which are involved in the phosphate solubilization is given in the following Table 20.

Table 20: Microorganisms and Sources of Phosphate

Sr.No.	*Microorganisms*	*Phosphate Sources*
	Bacteria	
1.	*Achromobacter* spp.	Lacithin (organic)
2.	*Aerobacter aerogenes*	Hexose monophosphatic ester (organic)
3.	*Alcaligenes* spp.	Phytin (organic)
4.	*Bacillus* sp., *B. megaterium, B. pulvifaciens*	Tricalcium phosphate (mineral)
5.	*B. circulans, B. mycoides, B. subtilis*	Calcium phosphate (mineral)
6.	*B. fluorescene, B. mesentericus*	Iron phosphate (mineral)
7.	*B. circulans*	Hydroxyapatite (mineral)
8.	*Brevibacterium* sp.	Calcium phosphate (organic)
9.	*Erwinia* sp.	Phenyl phosphate (organic)
10.	*Pseudomonas* spp. *P. putida*	Fluorapatiae (mineral)
11.	*P. liquifaciens*	Fluorapatiae (mineral)
12.	*P. calcis, P. rathonia*	Rock phosphate (mineral)

The fungi which bring insoluble phosphates in soil into soluble forms are given below:

Fungi

Aspergillus spp., *A. fumigatus*, *A. terreus*, *A. awamori*, *A. niger*, *A. flavus*, *Pencillum* spp., *P. lilacinum*, *Acrothecium* sp., *Cladosporium* sp., *Curvularia lunata*, *Candida* sp., *Fusarium* sp., *F. oxysporum*, *Humicola* sp., *Phythium* sp., *Phoma* sp., *Pseudogymnoascus* sp., *Sclerotium rolfsii*, *Streptomyces* sp., etc.

Selected bacteria are grown in Pikovskaiyas broth for 7-18 days at 28°C (±2°C) and mixed in suitable sterilized carrier (peat soil etc.). The mixture is cured for a week at 28°C (±2°C) in large trays covered with loosely fitted empty trays. Now the inoculant is ready to use. The inoculant is packed at the rate of 300g per plastic packet and stored at 15-20°C until use.

The inoculant can be used like *Rhizobium* inoculants. "Phosphobacterian" was first prepared in Russia by incorporating *B. megaterium* var. *phosphaticum*. Phosphobacterin is widely used in Russia and other East European countries for increasing the yield of crops by 5-10 per cent over corresponding control. In India, IARI used phosphobacteria for wheat, maize, rice, pigeon pea and berseem. The results are encouraging. The microorganisms *B. polymyxa*, *B. circulans*, *P. striata* and *Aspergillus awamori* are recently utilised for increasing the yield of crops like wheat and rice. Significant increase in the yield of above two crops is noted in India. *Pseudomonas* species are visualized as potent plant growth promoting rihzobacteria which are deleterious to seed borne rhizoshere inhabiting pathogens and therefore contributed to increase yield of potato in Netherlands.

Importance of Phosphate - Solubilizing Microorganisms

1. Recommended for all crops.
2. Can solubilize 20-30% of insoluble phosphate and increase the crop yield by 10-20%.
3. Microorganisms bring about insoluble soil phosphate into soluble forms by secreting several organic acids.

Rhizobium Inoculant

The bacteria in root nodules of legumes are called "Rhizobium". They are important agents in the fixation of atmospheric nitrogen. Chemical fixation of nitrogen came to know only several years after our understanding of the implications of biological nitrogen fixation.

It is known that man made fertilizers containing H_2, P_2O_5 and K_2 increase the yield of crops. It is also well known that the world's population, food production and fertilizer consumption have also increased considerably and biofertilizers is the need of day. The genus *Azotobactor* was well known in early part of 20th century in Russia and Poland as non-symbiotic nitrogen fixing bacteria. Azotobacterin was the trade product used for soil and seed treatment. It was effective on vegetable and cereal crops for increasing the yield.

The nodules on legume root are responsible for fixing atmospheric nitrogen. The leguminous plants are scattered in the family, leguminoseae. The family further divided into three sub families namely, Ceasalpinioideae, Mimosoidae and Papilionoideae. There are about 750 genera and about 18,000 to 19,000 species of leguminous plants. However, out of 20% examined plants of these families, 90% of Mimosoidae, 23% of Cesalpinipoideae and 97% of Papilionoideae shows root nodules. In addition, actinomycetes are also responsible for nitrogen fixation in some other families of plants. *Rhizobium* is widely scattered in the following families-

1. Azotobacteriaceae
2. Mycobacteriaceae
3. Myxobacteriaceae and
4. Psudomonadaceae.

Rhizobia are cultured on YEMA medium. They are able to produce acid or alkali on YEMA medium. The fast growing individuals are acid producers and slow growing are non acid producers. Slow growing rhizobia are believed to be the ancestors of rhizobia.

Fast growing rhizobia (acid producing) includes,

R. phaseoli,

R. trifolii,

R. leguminosarum and

R. meliloti

Slow growing rhizobia (non-acid producing) includes,

R. japonicum,

R. lupini and

Rhizobium sp.

Nitrogen fixation is an anaerobic process. Therefore, the nodule must have a mechanism to exclude oxygen from the bacteroiod. Bacteroid is the site of fixation. The leghaemoglobin present around bacteroid helps in the process of nitrogen fixation.

Rhizobium inoculation is tested on red gram (*Cajanus cajan*), gram (*Cicer arientinum*), Mung (*Vigna mungo*) broad bean and lentil. A significant increase in the yield of broad bean and lentil due to rhizobial inoclulation in the presence of superphosphate have been noted. Increase in the yield of crops like arhar, urid, soyabean, gram and lentil have been noted by legume inoculation with rhizobium. All India ICAR pulse project tested *Rhizobium* on chick pea, pigeon pea and mung at different states. The results are very encouraging and significant increase in the yield of all pulse crops tested have been noticed. Legumes are well known to leave behind some residual nitrogen in soil. Such residual nitrogen increases the yield of crops like wheat, rice and soyabean. By the application of rhizobium a good deal of money can be saved. Specially, such biofertilizers are useful to marginal farmers of India. However, the farmer should use quality tested inoculants on their farms. In this regard, small and large scale industry have significant role for uplifting the socio-economic status of marginal farmers by providing good service to them.

Rhizobium belongs to the family–Rhizobiaceae. There are three genera which produce Rhizobium.

1. *Azorhizobium*: It is used for stem inoculation.
2. *Bradyrhizobium*: It is useful against cow pea, gram, red gram, green gram, groundnut, etc.
3. *Rhizobium*: It is useful against pea, lentil, bean, lotus, berseem, etc.

Characters

It is unicellular with cell size less than 2 μ, wide, short to medium and rod shaped; and gram negative. It shows peritrichous flagella, is motile and gram negative. It is chaemoheterotrophic and symbiotic with legume, nitrogen fixed from atmosphere, respiration aerobic. *Rhizobium* grow fast while, *Bradyrhizobium* grow slow. Doubling time is 2-4 hours in fast growers and 6-12 hours in slow growers. Growth media is YEM/YEMA.

Contribution

It is applied to various crops in following dose-

1. Pigeon pea - 168-200 kg N/ha
2. Chick pea - 85-110 kg N/ha
3. Cow pea - 80-85 kg N/ha
4. Lentil - 90-100 kg N/ha
5. Groundnut - 50-60 kg N/ha
6. Green gram - 50-55 kg N/ha
7. Pea - 52-77 kg N/ha
8. Soyabean - 60-80 kg N/ha

Residual nitrogen benefits are recorded for the succeeding crops. It is recommended for legumes, pulses like cow pea, pigeon pea, gram, lentil, pea, green gram, etc. and oil seeds like soyabean and groundnut. It is also recommended for berseem and lucerne. It can increase the yield of crops from 10-35%. Some promising strains of *Rhizobium* used in India as biofertilizer are listed in Table 21.

Table 21: Promising Indian Strains of *Rhizobium* as Bio-fertilizer

Strains	*Crop*
IGR-6, NC-92	Groundnut
IGR-40	Groundnut
CC-1, F-4, BDN-A2	Pigeon pea
F-6, F-76, H-45	Gram
SB-103, SB-119	Soyabean
L-6, L-1	Lentil
RL(P)-114	Pea
M-10, M-11, GMBS-1	Green gram
DU-2, KU-1	Urid
DC-6, DC-2, DC-28	Cow pea

Azotobacter Inoculant

Mostly bacteria and blue-green algae are responsible for non-symbiotic nitrogen fixation in plants.

Free living nitrogen fixing bacteria are differentiate into following three groups.

1. Aerobic

2. Anaerobic and
3. Facultative anaerobic

The above classification is based on the growth and survival of the organisms in the presence or absence of oxygen.

Aerobic Bacteria

The aerobic bacteria have ability of fixing nitrogen belong to entirely different families. Aerobic bacteria have been isolated from different habitats like soil, freshwater, marine, animal and others. The aerobic bacterial species are scattered in the genera-

1. *Azotobactor*
2. *Azomonas*
3. *Azospirillum* (micro-aerophilic)
4. *Beijerinckia*
5. *Derxia* and
6. *Mycobacterium*

Anaerobic bacteria

These species are strictly anaerobic in fixing the nitrogen of the plants. Important anaerobic bacterial genera includes-

1. *Clostridium*
2. *Chlorobium*
3. *Chromatium* and
4. *Desulfovibrio*

Facultative Anaerobic Bacteria

The facultative anaerobic bacteria are visualised from the following genera.

1. *Bacillus*
2. *Enterobacter*
3. *Eschelrichia*
4. *Klebsiella*
5. *Rhodopseudomonas* and
6. *Rhodospirillum*

Azotobactor is heterotrophic bacteria which belongs to family-*Azotobaceriaceae* and depends on energy derived from the degradation of plant residues. In *Desulfovibrio* nitrogen fixation is possible by reduction of sulphates.

Diversity of Azotobacter

Azotobacter shows following species diversity.

1. *A. chroococcum* (soil inhabiting, brown black pigmentation)
2. *A. agilis* (water borne, green fluorescent pigmentation)
3. *A. vinelandii* (soil and water borne, with systs)
4. *A. beijerinkii* (soil inhabiting, yellow light brown pigmentation)
5. *A. insignis* (water inhabitant, syst less, grayish blue)
6. *A. macrocytogenes* (soil, systless, pink pigmentation)
7. *A. paspali* (soil, with syst)

The *Azotobacter* is characterised by following features. Cell size large ovoid, ranging from 2.0-7.0 x 1.0-2.5 μ, cell polymorphic, accumulate poly-β-hydroxybutyrate granules. It is gram negative chemohetrotrpophic and free living. It derives its carbon from mono, di and certain polysacharides and organic acids while, nitrogen through fixation, amino acids, NH_4^+ and NO_3^-. Its doubling time is 3 hours and grow on Ashby/Jensen's medium.

Contribution

1. Production of growth promoting substances such as vitamins of β-groups. Indole acetic acid and gibberellic acid.
2. Biological control of plant diseases by suppressing *Aspergillus* and *Fusarium*, etc.
3. 20-40 mg BNF/g of C-source in laboratory condition equivalent to 20-40 kg N/ha.
4. It is recommended for rice, wheat, millets, cotton, sunflower, mustard, vegetables and other cereals, etc.

Azospirillum

Beijerinck (1925) first described *Spirillum lipoferum* (=*Azospirillum*) as a nitrogen fixing bacteria. But, Becking (1963) and Bulow & Dobereiner (1975) emphasized on its application. The survey and review of literature indicates that *S. lipoferum* is a very common root and soil inhabiting nitrogen fixing bacteria in the tropical countries.

Azosprillum belongs to the family spirillaceae. There are four species reported under this genus namely,

1. *A. liproferum*
2. *A. brasilense,*
3. *A. amazonese* and
4. *A. halopraeferens/seropedicae*

Characters

Cell size curved rod like, 1 mm in diameter, or size and shape vary; accumulate poly-β-hydroxybutyric acid. It is gram negative with development of white pellicles 2-4 mm below the surface of NFB medium. It is associative and chemoheterotrophic in nature. It has sole carbon source from organic acids, L-arabinose, D-fructose, D-glucose, Sucrose, Pectin and D-gluconate. It has nitrogen source through fixation, amino acids, N_2, NH_4^+ and NO_3^-. It respires aerobically and micro aerobically. It has doubling time of one hour on ammonia containing media while, it take 5.5 to 7 hours on a malate containing semisolid medium. It has growth media of nitrogen free bromothymol blue (NFB).

A. lipoferum and *A. brasilense* may be differentiated by the following characters:

Sr.No.	*Medium*	*A. lipoferum*	*A. brasilense*
1.	Semisolid NFB medium with Glucose	Heavy growth on surface after 72 hours	No growth or pellicle much below surface after 72 hours
2.	Semisolid NFB medium with 0-0.005% yeast extract	No growth after several successive transfers	Normal growth

Contribution

1. 20-40 mg N fixed/g malate under laboratory condition equivalent to 20-40 kg N/ha.
2. Increases mineral and water uptake, root development, vegetative growth and crop yield.
3. It is effective against rice, millets, maize, wheat, sorghum, sugarcane, etc.

4. It increases the yield from 15-30% in the crops.
5. Recently, Reetha *et al.*, (2002) suggested that the combine effect of *Azospirillum* inoculation and foliar spraying of ACF (*Azospirillum* culture filterate) was more pronounced in augmenting the growth and yield of sunflower than the individual effect of them. The beneficial effect of *Azospirillum* is due to many factors such as nitrogen fixation, enhanced mineral uptake, production hormones and signal molecules. However, hormonal effects are the main mechanism in the plant growth promotion by *Azospirillum*.

Blue Green Algae (BGA)

Blue green algae is also called as "*Cyanobacteria*". The flooded rice eco-system is extremely complex, physiologically, physically, chemically and micro-biologically and well suited for habitation of blue green algae. Standing water in paddy field is favorable for the growth of blue green alage or "*Cyanobactria*". Blue green algae are scattered in several families like Nostocaceae, Scytonemataceae, Rivularicease, Stigonemataceae, Mastigocladaceae, Oscillattriaceae, Chlorococaceae and Oscikatoriaceae. From the fields of paddy more than 25 genera have been reported (Stewart, 1974). The list of some important and dominant genera found in the rice field is given below.

Genera	Species known in India
Anabaena	25
Anabaenopsis	3
Aulosira	8
Nostoc	25
Calothrix	29
Cylindrospermum	14
Gloecocapsa	-
Scytonema	49
Plectonema	11
Tolypothrix	19
Stigonema	12
Haplosiphon	10
Fischerella	(3)

Contd...

Genera	Species known in India
Mastigocladus	(1)
Westiella	(2)
Westiellopsis	(1)
Campylonema	-
Microchaete	-

Many filamentous blue green algae like *Nostoc*, *Anabaena*, etc have spherical cells and heterocysts. Heterosysts are spherical, oval cells with two layered walls. They may be located at intermediate or terminal position. Cyanobacterial cells are unicellular, colonial or filamentous, spherical or cylindrical in shape. Cyanobacteria are photolithotrophic and respire aerobically (heterosyst forming) or anaerobically (non heterosyst forming). They have C & N source from photosynthesis and atmosphere respectively and have doubling time of 20 hours.

Contribution

1. Fixation of about 20-30 kg N/ha in submerged paddy fields.
2. Production of Auxin, Gibberellic aicd, acetic acid and indole.
3. Effective in the paddy fields.
4. Increase the yield of crop from 15% to 20%.

The cost of algal material when produced on the farmers field is negligible but, commercially the cost for 10 kg/ha of algae may come to Rs. 30. An average increase of 300 kg/ha of grain is possible by algal inoculation (Subba Rao, 1995).

Azolla

Azolla in an aquatic fern that grows as a free floating hydrophyate in stagnant water. It is ubiquitous in distribution and very useful biofertilizer to rice crop. The fern *Azolla* always forms a green mat over water. The mat often becomes reddish due to the antocyanin pigments. Air spaces in the leaves of *Azolla* are occupied by a blue green algae, *Anabaena azollae*. *A. azollae* multiplies in the air spaces and remains there till the decay of the leaves. There is mutualistic association between *Azolla* and *Anabaena*. *Anabaena* fixes the nitrogen from air into nitrogenous compounds which accumulate in the air spaces in leaves. Later, released in water after the decomposition of leaves.

Azolla belongs to the family *Azollaceae* (fern) and *Anabaena* to Nostocaceae (symbiont). There are seven species of *Azolla* namely,

1. *A. caroliniana*
2. *A. nilotica*
3. *A. filculoides*
4. *A. mexicana*
5. *A. microphylla*
6. *A. pinnata* and
7. *A. rubra.*

Azolla was first applied as an organic input to rice in North Vietnam in 1957. Later, it has been tried in USA, Indonesia, Japan, Philippines and China (Tally, *et al.*, 1977). It is recommended for submerged rice field, within maximum termperature of 38°C.

Characters

Azolla	*Anabaena azollae*
1. Triangular or polygonal in shape	1. Barrel shaped, without sheath
2. Each leaf consists of dorsal and ventral lobe	2. Live in the cavities of dorsal lobe of *Azolla*.
3. Multiply mostly vegetatively	3. Multiply by binary fission.
4. C-source photosynthate from sunlight	4. supplied by *Azolla*.
5. N-source -N fixed by the symbiont	5. N-fixed from atmosphere
6. Doubling time 1-3 days (opt-cond), 5-7 days in field	6. 20 hours only

Limitations in the Use of *Azolla*

Azolla has limited use, only in flooded rice eco-system. However, *Azolla* is widely used in rice crop in several countries such as China, USA, Vietnam, Thailand, Indonesia, Philippines, India, etc. Symbiont *Azolla* increase the yield of crop from 10-25% in India.

In India, CRRI, Cuttack made significant contribution on *Azolla* through research pertaining to its use. Two methods of application of *Azolla* have been developed in India.

Life Cycle of Azolla (Schematic Representation).

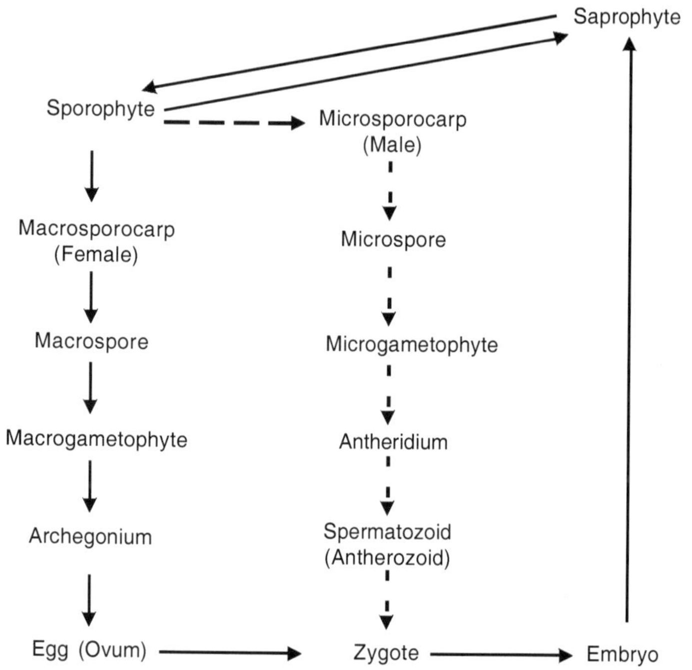

N.B. Broken arrows indicate stages where no transfer of lower symbiont takes place.

1. As a green manure by incorporating in field prior to rice planting and
2. By dual cropping with rice when the fern grows side by side with the main crop for some time.
3. Azolla is recommended for submerged paddy fields, within maximum temperature of 38°C. It has be observed that *Azolla* can increase the average yield of crop within the range of 15% to 20%.

Mycorrhiza

Mycorrhiza is the symbiotic association between fungal mycelia and roots or rhizomes of higher plants from pteridophyta, gymnosperms and angiosperms. Mycorrhiza increases the surface area of the root system for better absorption of nutrients from soil and keep plants healthy and fast growing. Considering the morphological and anotomical features the mycorrhiza are grouped into two types.

1. *Ectotrophic mycorrhiza* (Ectomycorrhiza)
2. *Endotrophic mycorrhiza* (Endomycorrhiza)

Ectotrophic Mycorrhiza

In this association, fungal mycelium forms a covering on the surface of roots. Root growth is stunted in infected plants. It becomes swollen and appears coralloid due to dichotomus branching.

The ectomycorrhizal fungi belongs to Basidiomycetes and Ascomycetes and few to endogonaceae. The higher plants belonging to the families, Betulaceae, Diptherocarpaceae, Ceasalpiniaceae, Pinaceae, Fagaceae and Saliceae have the ectomycorrhizal associations. The genera like *Eucalyptus, Tilia, Arbutus* and *Casuarina* also shows ectomycorrhizal assocation.

Endotrophic Mycorrhiza

In this case the fungal hypae are present inside the root. The hypae grow in between and within the cells of cortex of the roots. On the root surface a very little growth of mycelium is seen. In the endomycorrhizal association, plants of most families of angiosperms, conifers (except pianceae), pteridophytes and bryophytes are involved. Many species of the following genera of orchids depends on endomycorrhizal fungi for their growth.

Orchid Genera	*Fungi Genera*
Neottia	*Fomes*
Limodorum	*Corticium*
Vanilla	*Rhizoctonia*
Epipogon	

Some endomycorrhizae are called as VAM (Vascular-arbuscular mycorrhiza) since they possess spherical structures like vesicle and arbuscules. The arbuscules help in the transfer of nutrients from soil into the root system. VAM species are scattered in five genera. Their classification is based on the spore morphology. Many graminaceous plants and legumes are highly susceptible to VAM.

Important species of five genera are listed below with their specialities.

1. *Glomus mosseae*
2. *G. fasciculatum*
3. *Gigaspora nigra*
4. *Acaulospora scrobiculata*
5. *Sclerocystis clavispora*
6. *Endogone increseta*

Glomus microcarpus have spores of 50 μ diameter and *G. mosseaor* and *Gigaspora cladospora* have 100-200 μ diameter. The spores of *Glomus* spp. are single stalked, while, *Gigaspora* shows bulbus stalked spores. However, the spores of *Acaulospora* are always sessile. VAM are obligate symbiot. They have not been isolated in pure culture. They can be maintained on growing plants.

Contribution

1. Enhances uptake of P, Zn, S and water.
2. Increases growth and yield of crops.
3. Promotes uniform growth of crops.
4. Increase resistance in root against diseases.
5. Reduces stunting on fumigated soil.
6. Helps drought stressed plants.
7. It is useful against forest trees, forage grasses, legumes, cereals, etc.

Green Manure

The compost which is prepared from freshly collected green materials of sherbs, herbs and trees is called as green manure. The plants are uprooted and ploughed back in the fields at the time of flowering. Green manuring helps to increase the nitrogen content of the soil. Green manuring is specially useful in flooded rice eco-system. The flooded rice soil provide a congenial microhabitat for a variety of nitrogen fixing micro-organisms. In such system rice can grow year after year with minimum but constant yield, without additional chemical fertilizers. The rice crop use, only 20-40% of added nitrogen. As like blue green algae, Azolla and microorganisms, "legume green manure", is also an important source of biologically fixing nitrogen.

The fresh green matter of legumes can be collected from plant species grown specially for this purpose and from naturally grown plants. The leaves of shrubs, herbs and trees of legumes are collected, transported and applied to rice fields prior to seedling transportation.

The genera *Astragalus, Aeschynomene, Crotalaria, Indigofera, Lathyrus, Pongamia, Tephrosia, Trifolium* and *Sesbania* are maily used as green manure crops in wetland rice farming (Allen & Allen, 1981).

Arachis, Cajanus, Calopogonium, Atylosia, Canavalia, Cassia, Cullen, Cochlianthus, Derris, Dipogon, Lablab, Dolichos, Dunbaria, Mimosa, Mucuna, Eruthrina Flemingia, Gliricidia, Kummerowia, Lespedeza, Leucanea, Lupinus, Mastersia, Melilotus, Pisum, Ormocarpum, Pachyrhizus, Parkia, Phaseolus, Pteroloma, Psophocarpus, Peueraria,Shuteria, Stylosanthes, Tetragonolobus, Trigonella, Uraria, Vigna, Vicia, etc. are important genera for green manuring to plantation crops and *Aeschynomene, Neptunia* and *Sebania* are important genera among the legumes which are known to bear nodules on stems. The genus *Aeschynomene* contains 150-250 species of herbs to small and medium sized trees. In India, *A. indica* and *A. aspera* species are more familiar.

The genus *Sesbania* contain 170 species. *S.rostrata* is cosmopolitan species and useful in rice agro-eco-system. Indian green manuring is well known to the world. Specially, South Indian farmers practiced the green manuring since long back. Sunhemp (*Crotalaria juncea*), Dhanicha (*Sesbania aculeta bispinosa*), Pillepesara (*Phaseolus trilobus*), Mung (*Vigna mungo*), Cowpea (*V. unguiculata*) Gaur (*Cyamopsis tetragonoloba*), Sengi (*Melilotus alba*), Berseem (*Trifolium alexandrinum*) and Khesari (*Lathyrus sativus*) are practiced in rice field in India.

Indian green manuring is well known to world. South Indian farmers practiced the green manuring since long back. Sunhemp (*Crotalaria juncea*), Dhaincha (*Sesbania aculeta bispinosa*), Pillipesara (*Phaseolus trilobus*), Mung (*Vigna mungo*), Cowpea (*V. uniguiculata*), Gaur (*Cyamopsis tetragonoloba*), Sengi (*Melilotus alba*), Berseem (*Trifolium alexandrinum*), and Khesari (*Lathrys setivus*) are practiced in rice field in India.

In Southern India, leaves of *Cassia auriculata, Calotropis gigantea, Jatropha gossipifolia, J. glandulifera, Tephrosia purpuria, T. candida,*

Cassia tora, Ipomea carnea, Melia azadirachta, etc. are used as leaf green manure. In northern parts of India, *S. aculeta* is preferred in rice field. It is sown after wheat in April, the biomass turned into soil in standing water after 8 weeks of growth followed by transplantation of seedlings of rice. The application of green manure of sunhemp, cowpea and Dhanicha increases the yield of rice crop and also nitrogen content of the soil.

Organic Matter and Composting

The primary sources of organic matter in soil are dead and decaying parts of plants and animals. The organic matter, further consists of soluble sugars, amino acids and organic acids and insolubles like celluloses and legnin. Fats, oils, waxes, resins, pigments, minerals, proteins, etc. are also other components of organic matter. The above mentioned matter is given out by microorganisms in soil. As a rule microorganisms feed readily on soluble resources and slowly on insoluble matter.

Through assimilation, the carbon content of the protoplasm of microorganisms is derived mainly from organic substrates. From organic matter entire carbon is not assimilated into the microbial protoplasm. However, the assimilation of carbon depends upon the nature of microflora and the oxygen status of the soil. Immobilization process is also a very prominent process in organic matter composting under which nutrients such as N, P, K and S are taken up by microbial cells at the time of carbon assimilation. Mineralization is the conservation of organic complexes of an element to its inorganic state. It is also related with organic matter and composting.

Fungi releases less CO_2 and incorporate more carbon in their tissues. But, in some aerobic bacteria it is reverse process. From organic matter, methane (CH_4) and hydrogen (H_2) are evolved under anaerobic condition in incomplete degrading organic matter. While, several intermediate organic acids, CH_4, H_2 and CO_2 are generated in waterlogged situations. From organic acids mainly acetic, formic, butyric and lactic acids are released. In addition, lactic and succinic acids are released. In addition, lactic and succinic acids are also released but in lesser quantities. O_2 uptake, loss of weight, magnitude of CO_2 formation, etc. are important parameters for measuring organic matter decomposition.

When fresh plant residues reach soil, immobilization became slow, however, mineralization gradually increases depending upon temperature, water content, oxygen supply, hydrogen ion concentration, inorganic nutrients, the nature of microflora, nitrogen content of residues and C/N ratio of the organic matter.

Leguminous plant residues encourage the growth of microorganism by supplying readily available nitrogen and accelerate decomposition. In nitrogen deficient cereals, straw residues, additional ammonia or nitrate accelerate decomposition and microbial activity.

The branches of older trees contain lignin and residues of younger plants contain soluble nutrients. Hence, younger plant residues are metabolized more easily than residues of older trees. In acidic soil, filamentous fungi degrades cellulose. Likely, in neutral and alkaline pH levels, many microorganisms act on cellulose. *Chaetomium* and *Trichoderma* are good examples of fungi and *Cellulomonas* and *Clostridium* of bacteria.

Humus contain amino acids, pyrines, pyrimidines, amino sugars, methyl sugar, pentoses, hexoses, sugar alcohols, aliphatic acids, uric acids and aromatic substances. The age of humus can be identified by radio carbon dating techniques. According to Williams (1971) certain fungi like *Aspergillus* and *Penicilium* and some actinomycetes produce dark humus which contain amino acids, peptides and polyphenols.

According to the reports of ICAR addition of farm yard manure increases the number of root nodules of leguminous plants and also the number of *Azotobacter* cells in soil. In pea (*Pisum sativum*), Dhanicha (*Sesbania aculeata*) and berseem (*Trifolium alexandrinum*) the number of nodules, nodule weight and yield have been increased due to application of specific calcium and sodium humate. Likely, in soyabean, nodule number and dry weight yield increased due to use of pelleted seeds with calcium humate when grown in a saline and alkaline soil.

In India, Fowler (1930), Howard (1935), Acharya (1939, 1949), etc. contributed on composting. They described advantages and disadvantages of composting practices in India, specially in rural areas. It is noted that rural method or native method of composting is crude and unscientific.

The scientific methods designed in India includes,

1. Indore method (aerobic), and
2. Bangalore method.

Indore Method

A pit of 3 feet deep and 6-8 feet wide is dug near a cattle shed. The place must be free from water logging. A cattle bed is prepared with the plant residues cut into sherbs and used for cattle in shed. After use, the bed with wet excreta is removed and heaped in a layer of 3-5 meter in the pit. A sufficient quantity of cattle dung is then spread over this layer. Later, water is sprinkled to maintain moisture. Above sequential layering is followed till the residues reach one feet above the ground level. The excess rains may be avoided by erecting shed over the pit. The residues are turned every two weeks, thus, within 16 weeks the compost will be ready for use.

Bangalore Method

In Bangalore methods, pits are similar to Indore method. The preparation of heaps is also similar to Indore method except that each heap in a pit is sealed with plaster of mud. This increases the temperature inside the pit by an anorobic fermentation and thus nitrogen-rich compost is formed. For the complete process of composting it takes 32 weeks. This method was proposed by ICAR in 1964. The use of additional microorganisms accelerate composting process. The celluloytic microorganisms used in the process are *Trichoderma viride, Aspergillus niger, A. terreus, Chaetomium abuanse, Myrothecium roridum* and *Cellulomonas*, etc.

Dethier, V.G. (1947). Chemical insect attractants and repellents. Blakiston, Philadelphia, p. 289.

Dutta, B.L. (1962). *Indian Potato J.*, **4**: 27.

FAO (1966-72). Reports of FAO panel of experts on Integrated Pest Control, Rome.

Fuerst, E.P. and Norman, M.A. (1991). Interactions of herbicides with photosynthetic electron transport. *Weed. Sci.*, **39**: 458-464.

Flower, G.J. (1930). Recent experiments on the preparation of organic manure: A review. *Agric. Livestock Indian*, 7: 711-717.

Gangawane, L.V. and Despande, J. (1985). Pesticides of crop plants in India. Sahyog Publ. Aurangabad. pp. 1-300.

Gilbert, L.I. (1976). The Juvenile Hormones Plenum.

Grewal, J.S. and Singh, R.P. (1965). *Insect phytopathology*, **18**: 226-228.

Gronwald, J.W. (1991). Lipid biosynthesis inhibitors. *Weed Sci.*, **39**: 435-439.

Gruzdyev, G.S. (1983). The chemical protection of plants. Mir Publishers, Moscow.

Hall, S.A., Green, N. and Beroza, J.M. (1957). *J. Agr. Food Chem.*, **5**: 663.

Howard, A. (1935). The waste products in agriculture. *J.R. Soc. Arts*, **84**: 120.

Howell, D.E. and L.H. Goodhue (1955). *Soap Sanit Chemicals*, **31**: 181.

Hutton, K.E. (1961). "Bitter rot of apples". *Agric. Gaz. N.S.W.*, **72**: 319-322.

ICAR (1964). Hand book of manures and fertilizers. ICAR, New Delhi.

Ingram, J. (1963). *Expt. Hort.*, **8**: 38-47.

Inoue, Y. and Ohno, M. (1961). Insect attractants. *Kagaku No Ryoiki*, **15**: 823-832.

Jacob, S. and Sheila, M.K. (1994). Studies on the antifeedant activity of some plant products against the leaf caterpillar. *Selepa docilis* Butl. on brinjal and wooly bear, *Pericallia ricini* F. on caster. *Indian J. Ent.*, **56**: 276-279.

Jacobson, M. (1963). Recent progress in the chemistry of insect attractants. *Advan. Chem. Sci.*, **41**: 1-10.

Jacobson, M. (1972). Insect sex pheromones. Academic Press, N.Y. pp. 382.

Jacobson, M. and Bezona, M. (1964). Insect attractants. *Sci. Am.*, **211**: 20-27.

Knipling, E.F. (1968). The potential role of sterility in pest control. In: *Principles of insect chemosterilants*. NHP Co.

Kinpling, E.F. (1955). *J. Econ. Entomol.*, **48**: 459-462.

Kinpling, E.F. (1959). *Science N.Y.*, **10**: 902-904.

Kinpling, E.F. (1960). *J. Econ. Entomol.*, **53**: 415-420.

Lindquist, A.W. (1955). *J. Econ. Entomol.*, **48**: 467-469.

Lindquist, A.W. (1961). *Pest Control*, **29**: 9-40.

Lindsay, E. and Craig, R. (1942). The distribution of radiophosphorus in wax moth, meal worm, cockroach and firebrat. *Ann. ent. Soc. Am.*, **35**: 50-56.

Marsh, R.W. (1972). Systemic fungicides. WC & Sons Ltd., London.

Mayer, K. (1952). *Pharmaziee*, **7**: 150-157.

Mishra, P.K. and Singh, R.P. (1992). Antifeedant efficacy of neem (*Azadirachta Indica*) seed kernel, seed coat and fallen leaves extracts against desert locust *Schistocerca gregaria* F. *Indian J. Ent.*, **54**: 89-96.

Motsara, M.R. and Bhattacharya, P. and Beena Shrivastva (1995). Biofertilizer technology, marketing and use. EDCO, New Delhi. pp. 1-184.

Munakata K., Saito, T., Ogawa, S. and Ishil, S.O. (1959). *Bull. Agri. Chem. Soc.*, Japan. **232**: 64-65.

Nayar, J.K. and Fraenkel, G. (1963). *Ann. ent. Soc. Am.* **56**: 174-178.

Nene, Y.L. and Agarwal, V.K. (1969). *Pesticides*, 3: 15-17.

Nene, Y.L. (1970). Fungicides in plant disease control. Oxford & IBH Publ. Co., New Delhi.

Nene, Y.L. and Thapliyal, P.N. (1993). Fungicides in plant disease control, pp. 1-647.

Novak, V.J.A. (1975). Insect hormones. Chapman & Hall, London.

O'Brien, R.D. (1967). Insecticide action and metabolism. Academic Press.

Oswal, K.P. and Bhide, V.P. (1972). *Indian J. Expt. Biol.*, **10**: 153.

Parker, D.R. (1959). Dominant lethal mutation in irradiated oocytes. Uni. Taxas Publ. 5914: 113-127.

Pradhan, S. (1967). Insect pests of crops. NBT.

Quarishi, M.S. (1977). Biochemical insect control. J.W. & Sons.

Rao, V.S. (2000). Principles of weed science. Oxford & IBH, New Delhi, pp. 1-231.

Remold, H. (1963). *Nature*, **198**: 764-768.

Riley, E.A. (1955). "Tobacco Anthracnose". *Trop. Agriculture Trin.*, **32**: 150-155.

Sathe, T.V. (1988). Sericulture crop protection. Asawari Pub. Osbad.

Sathe, T.V. and Bhoje, D.K. (2000). Biological pest control. Daya Publishing House, Delhi.

Sathe, T.V. and Jadhav, A.D. (2001). Sericulture and Pest Management. Daya Publishing House, Delhi

Sathe, T.V. and Girhe, B.E. (2002). Mosquitoes and Diseases. Daya Publishing House, Delhi.

Sathe, T.V. and Margaj, G.S. (2001). Cotton Pests and Biocontrol Agents. Daya Publishing House, Delhi.

Saxena, B.P. and Koul, O. (1978). Utilization of essential oils for insect control. *Indian Perfumer*, **22**: 139-149.

Sengupta, K., Govindalah and Pradip Kumar (1991). Diseases and pests of mulberry and their control. CSR & TI, Mysore.

Singh, B. and Gupta, S.L. (1953). Deptt. Agric. U.P. (India), *Tech. Bull.*, **12**: 1-16.

Sharvelle, E.G. (1960). The nature and uses of modern fungicides. Burges Publ. Co., USA.

Shetty, S.V.R. and A.N. Rao (1981). Weed management studies in sorghum/pigeonpea and pearl millet/groundnut intercrop systems. Some observations. *Proc. Internat. Workshop on Intercropping*, 10-13 Jan., 1979, ICRIST, Hyderabad, pp. 238-248.

Siddiqui, J.I. (1988). Perception of sex pheromone in *Dicrisia obliqua* Walker (Lepidoptera-Arctiidae). *Indian J. Ent.*, **50**: 343-349.

Stewart, W.D.P. (1974). Blue green algae. In: Biology of nitrogen fixation. Ed. A. QNH Publ. Co., Amsterdam, pp. 202-287

Subba Rao, N.S. (1995). Biofertilizers in Agriculture. Oxford & IBH.

Sundaran, N.V. (1964). *Madras Agric. J.*, **51**: 92.

Swanton, C.J. and Weise, S.F. (1991). Integrated weed management: The rational and approach. *Weed Technol.*, **5**: 657-663.

Talley, S.N., Talley, B.J. and Rains, D.W. (1977). Nitrogen fixation by *Azolla* in rice fields. In genetic engineering for nitrogen fixation. Ed. A. Hollaender, Plenum Press, New York, pp. 259-281.

Tamaki, Y. (1985). Sex pheromones, D.P. New York, pp. 142-192.

Tandon, I.N. (1961). *Indian phytopathology*, **14**: 102.

Tandon, I.N. (1963). *Indian phytopathology*, **16**: 44-47.

Turner, R.B. (1968). "Chemistry of insect chemosterilants" In: Principles of insect chemosterilization. ACC, New York.

Torgeson, D.C. (1969). Fungicides: An advanced treatise-II.

Varadarajan, B.S. and J.S. Patel (1946). *Indian J. Agric. Sci.*, **16**: 193-206.

Verma, S.K. (1990). Efficacy of antifeedant triphyltin acetate against the Vth instar larvae of *Amsacta moorei* Butler. *Indian J. Ent.*, **52**: 521-524.

Vyas, S.C. (1984). Systemic fungicides. Tata Mc-Graw Hill, New Delhi.

Whiting, A.R. and Bostion, C.H. (1931). The effect of X-radiation of larvae in *Habrobracon*. *Genetics*, **16**: 659-680.

Wilkinson, C.E. (1976). Insecticides Biochemistry and Physiology. Plenum.

Williams, C.M. (1967). Third generation pesticides. *Scientific American*, **217**: 13-17.

Yamamoto, I. (1970). Mode of action of purethroids, nicotinoids and rotenoids. *Ann. Rev. Entomol.*, **16**: 257-272.

INDEX

B

D

T